Henry Leffmann, William Beam

Analysis of Milk and Milk Products

Henry Leffmann, William Beam

Analysis of Milk and Milk Products

ISBN/EAN: 9783337146047

Printed in Europe, USA, Canada, Australia, Japan

Cover: Foto ©berggeist007 / pixelio.de

More available books at **www.hansebooks.com**

ANALYSIS

OF

MILK AND MILK PRODUCTS

BY

HENRY LEFFMANN, M. D., PH. D.,

PROFESSOR OF CHEMISTRY IN THE WOMAN'S MEDICAL COLLEGE OF PENNSYLVANIA,
IN THE PENNSYLVANIA COLLEGE OF DENTAL SURGERY AND IN THE
WAGNER FREE INSTITUTE OF SCIENCE; PATHOLOGICAL
CHEMIST TO THE JEFFERSON MEDICAL
COLLEGE HOSPITAL,

AND

WILLIAM BEAM, M. A., M. D.,

FORMERLY CHIEF CHEMIST B. & O. R. R.

PHILADELPHIA:
P. BLAKISTON, SON & CO.,
1012 WALNUT STREET.
1893.

PRINTED BY E. F. GREATHEAD
909 SANSOM STREET
PHILADELPHIA

PREFACE.

The analysis of dairy products has been the subject of so much exact investigation within the last few years, that most of the earlier processes have been abandoned, since it has been shown that the data derived from them are inaccurate. In the present work we have had in view the object of providing not only methods suitable for professional chemists, but also such as may be safely employed by dairymen and others unskilled in general analytical work. To aid in securing uniformity, we have described some of the published methods of the Association of Official Agricultural Chemists, selecting those that we find to be satisfactory.

We have drawn freely from the valuable material contributed to *The Analyst* and these references will show how large a part of the advance in the analytical practice in this field is, due to the industry and ingenuity of the members of the Society of Public Analysts. We have constructed a table, based on the formula of Hehner and Richmond, arranged to suit the method of calculating total solids from the fat and specific gravity. Several other tables, not easily accessible, have been added, the insertion of which will contribute to the usefulness of the work.

715 Walnut St., Phila.   H. L.
    July, 1893.   W. B.

CONTENTS.

NATURE AND COMPOSITION OF MILK.
    Formation and Ingredients of Cow's Milk.—Colostrum—Milk of Various Animals—Properties and Decompositions of Milk—Skimmed Milk—Buttermilk, ... 10-14

ANALYTICAL PROCESSES.
    Specific Gravity—Total Solids—Ash—Fat—Proteids—Sugar—Milk Adulterants, . . . 15-46

DATA FOR MILK INSPECTION.
    Variations in Composition—Sanitary Relations, 47-59

MILK PRODUCTS.
    Condensed Milk—Butter—Cheese, . 60-78

APPENDIX.
    Table for Correcting Specific Gravity—
    Table for Calculating Total Solids, 79-89

NATURE AND COMPOSITION OF MILK.

Milk, by which term is to be understood the nutritive secretion of nursing mammalia, consists of water holding fat in suspension, and nitrogenous, saccharine

ERRATA

Page 10, line 3, for ".035 to .050" read "0.35 to 0.50."
Page 65, line 10, insert a quotation mark after word "*ash*."

eration of the epithelial cells lining the ducts of the mammary gland. It occurs in the form of minute globules from .0015 mm. to .005 mm. in diameter, under conditions which prevent spontaneous coalescence. Films of proteid matter are also abundant. A form of sugar, isomeric with cane-sugar, and called lactose, is present.

NATURE AND COMPOSITION OF MILK.

Milk, by which term is to be understood the nutritive secretion of nursing mammalia, consists of water holding fat in suspension, and nitrogenous, saccharine and mineral matters in solution. Cow's milk, being of the greatest importance to the analyst, will receive the largest share of attention, and will be understood to be meant in all cases, unless otherwise stated.

With rare exceptions, the secretion of milk takes place only as a result of pregnancy and delivery at term, and continues for a variable period. The chemistry of its formation is not entirely understood. The organic ingredients do not exist in appreciable quantities in the blood, and must, therefore, be elaborated by specific secretory action. The fat has been regarded by many authorities as resulting from the fatty degeneration of the epithelial cells lining the ducts of the mammary gland. It occurs in the form of minute globules from .0015 mm. to .005 mm. in diameter, under conditions which prevent spontaneous coalescence. Films of proteid matter are also abundant. A form of sugar, isomeric with cane-sugar, and called lactose, is present.

The proteids of milk consist largely of casein. There is also some albumin (the form called lactalbumin) from .035 to .050 per cent. being usually present, together with traces of globulin. The albumin is coagulated by boiling. The condition in which the casein exists is still obscure, but the results of recent investigation render it probable that it is in combination with the calcium phosphate. Acids break up this combination, and precipitate the casein in an insoluble form. Heat alone does not produce this effect. The form in which the casein exists prior to coagulation is now frequently called "caseinogen".

Citric acid is a normal constituent of the milk of various animals. In human milk, the quantity is about 0.5 gram to the liter, in cow's milk from 1 to 1.5 grams. It is not dependent on the citric acid present in the food.

Minute amounts of nitrogenous bases and a starch converting enzyme also occur.

The ash of milk has the following composition:

Ca	.113	per cent. (of milk)
Mg	.0126	" "
Fe	.0002	" "
K	.146	" "
Na	.082	" "
PO_4	.263	" "
SO_4	traces	" "
Cl	.169	" "
CO_3	.020	" "

Colostrum.—The milk secreted in the early stage of lactation is rich in proteids (20 per cent. or more) due probably to the incomplete transformation of the epithelial lining of the ducts. A protoplasmic structure known as the colostrum corpuscle is found at this period. The large proportion of proteids is often sufficient to cause the liquid to coagulate on boiling. The following is an average composition of colostrum:

Fat	3.5
Proteids	12.7
Sugar	4.2
Ash	0.8
	21.2

The following table is a compilation of published statements of the composition of various milks.

	Human	Cow	Porpoise	Mare	Cjamoose	Goat	Ass	Ewe	Sow
Specific Gravity				about 1000.0	1034.9	1035.4	1032.9		
Fat	3.5	3.9	45.80	1.09	5.56	4.3	1.6	6.8	4.8
Sugar	6.5	4.7	1.33	6.65	5.41	4.0	6.1	4.8	3.4
Proteids	1.6	3.7	11.19	1.89	3.86	4.6	2.2	6.3	1.3
Ash	0.25	0.7	0.57	0.31	1.03	0.6	0.5	0.8	0.9
Tot'l Solids	11.80	13.0	58.89	9.94	15.86	13.5	10.4	18.7	15.4
Analyst			F. Purdle	Vieth	Pappel & Richmond				

Normal milk is an opaque, white or yellowish-white fluid, with an odor recalling that of the animal, and a faint sweet taste. The opacity is due partly to the fat globules, but when these are entirely removed the liquid does not become transparent. The reaction of freshly drawn milk is amphoteric, that is, it turns red litmus paper blue and blue litmus paper red. The specific gravity varies between 1028 and 1035. It undergoes a gradual augmentation for a considerable time after the sample has been drawn. The increase may amount to two units. The specific gravity becomes stationary in about five hours, if the milk be maintained at a temperature below 60° F., but at a higher temperature it may require twenty-four hours to acquire constancy. The change is not dependent on the escape of gases, and is believed to be due to some molecular modification of the casein.

Unless collected with special care and under conditions of extreme cleanliness, milk always contains bacteria and animal matter of an offensive character, such as epithelium, blood and pus cells, particles of feces and soil. Many minute organisms, especially bacteria, propagate with great rapidity in milk and produce changes in its composition. Some specific organisms, such as the *Spirillum choleræ*, multiply to only a limited extent in ordinary milk, being hampered by the bacteria normally present, but when introduced into sterilized milk increase with great rapidity.

At ordinary temperature milk soon undergoes decomposition under the influence of the microorganisms present, by which the milk sugar is converted principally into lactic acid, and the proteids partly decomposed and partly coagulated. The liquid becomes sour and the fat is enclosed in the coagulated casein.

In the initial stages of decomposition the proteids frequently undergo transformations into highly poisonous benzene derivatives, among which diazobenzene, commonly known as tyrotoxicon, is the most important. This body is the cause of the violent poisonous effects occasionally produced by ice cream and other articles of food into the preparation of which milk enters.

Boiling produces coagulation of the albumin, some alteration of the sugar, and developes a greater facility of coalescence on the part of the fat globules. Microbes and enzymes are destroyed. The scum which appears on the surface of boiling milk is composed largely of casein. Its formation is due probably to the more rapid evaporation at the surface of the liquid.

Precipitation of the casein in the form of a curd, enclosing the fat, occurs promptly on the addition of rennet or free acids.

Partial freezing produces a concentration of the milk solids in the part remaining liquid, while the solid portion is deficient in them. The normal condition can, therefore, be restored only by thawing the entire mass, and mixing thoroughly.

When milk is allowed to stand, some of the fat rises gradually, and forms a rich layer, constituting cream. The proportion of cream depends on several conditions. The amount formed in a given time cannot be taken as a measure of the richness of the milk. Water added to milk causes a more rapid separation of the cream. When milk is subjected to centrifugal action, a much larger proportion of cream is obtained, practically all of the fat being removed. The following figures given by D'Hout as averages, show the effect of the centrifugal action:

	Whole milk	Skimmed milk	Cream
Specific gravity	1032	1034	1015
Total solids	14.10	9.6	26.98
Sugar	4.70	5.05	3.32
Casein	3.50	3.62	2.02
Ash	0.79	0.78	0.58
Fat	5.05	0.02	21.95

Buttermilk is the residue after removal of the butter by churning. Vieth gives the following figures:

Total solids	Fat	Solids not fat	Ash
9.03	0.63	8.40	0.70
8.02	0.65	7.37	1.29
10.70	0.54	10.16	0.82

ANALYTICAL PROCESSES.

SPECIFIC GRAVITY.

Determinations of specific gravity of milk are understood to be taken at the temperature of 60°F., and samples at temperatures different from this should be brought to it. If at a few degrees above or below 60° it will suffice to take the gravity at once, and obtain the correct figures by reference to Table A.

The specific gravity of normal milk varies between 1028 and 1035, being less in porportion as the fat is greater. A milk, the specific gravity of which has been raised by the abstraction of fat (skimming) can be restored to the original specific gravity by the addition of water, so that this figure taken alone cannot be relied upon as an index of the character of the sample. Taken in conjunction with the figure for fat or for total solids, it is of the greatest value, and should be ascertained as a check on the results furnished by other determinations.

Air bubbles are held rather tenaciously by milk and care must, therefore, be taken in mixing the samples preparatory to taking the density, to avoid as far as possible the enclosure of air, and to allow sufficient time for the escape of any bubbles that may be present.

ANALYTICAL PROCESSES.

The simplest method of determining specific gravity is by the *lactodensimeter*, a delicate and accurately graduated hydrometer. The instrument must be immersed carefully so as not to wet the stem above the point at which it will rest. The reading should be made at the actual level of the liquid, and not at the point to which it is drawn by adhesion to the glass.

The indications furnished by the lactodensimeter are sufficiently accurate for most purposes, but its employment necessitates a considerable amount of the sample.

More accurate determination can be made by the *Westphal balance.* This is a delicate steelyard with a counterpoised plummet, displacing 5 c. c. The plummet being immersed in the milk the equilibrium is restored by weights, the value of which can be directly expressed in figures for the specific gravity.

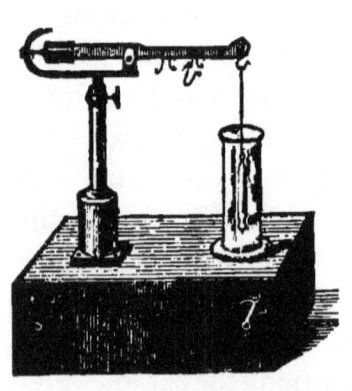

The principal of the Westphal balance may be applied by means of the ordinary analytical balance and a plummet. The latter may conveniently consist of a short thermometer, or a thick glass rod, having a bulk of from 5 to 10 c. c. It is suspended from the hook of the balance by a fine platinum wire and its weight ascertained. It is then immersed in distilled water at 60°F. and the loss in weight noted. The figure so obtained is the weight of a bulk of water equal to that of

SPECIFIC GRAVITY.

the plummet. This having been determined, the specific gravity of a milk may be found by immersing the plummet in it, and noting the loss in weight, which, divided by the loss in pure water, gives the specific gravity.

The *pyknometer* or *specific gravity bottle* furnishes a means of accurately determining specific gravity and is especially suitable when only a small amount of liquid is available. It consists of a small flask provided with a finely perforated glass stopper. The flask is weighed, first alone, then filled with water at 60°F., then with the milk at the same temperture. In filling the flask the liquid is first brought to the proper temperature, the bottle completely filled, the stopper inserted and the excess, forced out through the perforation and around the sides of the stopper, removed by bibulous paper. The weight of milk divided by that of an equal bulk of water gives the specific gravity.

TOTAL SOLIDS.

These are determined by evaporation in a shallow, flat-bottomed platinum or porcelain dish from 7 to 8 cm. in diameter. The milk must be spread evenly in a thin layer. If the ash is also to be determined, about five grams should be accurately weighed in the dish, evaporated rapidly to apparent dryness over the water-bath and the heating continued in the water-oven until the weight becomes practically constant, which will require about three hours. When the ash is not to be determined, it is more convenient to follow the method suggested by Richmond, using 1 to 2 grams, accurately weighed. The drying can be completed in about one and a half hours.

Richmond has pointed out that if the evaporation be slow, some decomposition occurs, and the residue will be brown, but that if the larger portion of the water be evaporated quickly, a white residue is obtained. He suggests the following method for use when a higher degree of accuracy is desired:

About 3 grams of abestos of the best quality are placed in a platinum dish, ignited in a muffle and weighed. By simple ignition over the Bunsen burner the combined water is not always lost. 5 grams of milk are added and dried on the water-bath for about two hours. The residue is then left in the air or water-oven at about 208°F. for twelve hours or more (usually over night). At the end of this time an absolutely constant weight is obtained.

The residue serves well for the determination of the ash.

Where rigid accuracy is not essential, it will suffice to measure the portion of milk taken for this and other determinations. Vieth uses a pipette graduated to deliver 5 grams and finds that, working with whole and skimmed milk, under the ordinary variations of temperture, the error will not exceed 0.1 on the total solids and is less on the fat.

A better plan is to use a 5 c. c. pipette and to wash out that which adheres to the glass with a little water. The specific gravity of the milk being known, the amount taken can be calculated. The milk should be as near 60°F. as possible.

ASH.

The residue from the determination of total solids is heated cautiously over the Bunsen burner, until a white ash is left. The result obtained in this manner is apt to be slightly low from loss of sodium chloride. This may be avoided by heating the residue sufficiently to char it, extracting the soluble matter with a few c. c. of water, and filtering (using paper extracted with hydrofluoric acid.) The filter is added to the residue, the whole ashed, the filtrate then added and the liquid evaporated carefully to dryness. The ash of normal milk is about 0.7 per cent, and faintly alkaline; if the milk be watered the ash will be less. It will be seen by the detailed statement of composition that it is practically free from sulfates, and hence a quantitive determination of these may detect the adulteration with water containing sulfates. For this purpose, not less than 100 c. c. should be taken. A marked degree of alkalinity and effervescence with hydrochloric acid, will suggest the addition of a carbonate.

FAT.

The method introduced by Wanklyn for the determination of fat by extracting it with ether from the total solid residue, has been found to give results 0.5 per cent or more below the correct figure, and is therefore not described.

Adams' Method.—This consists essentially in spreading the milk over absorbent paper, drying, and extracting the fat. By this means the milk is distributed in an extremely thin layer, and by a selective action of the paper, the larger portion of the fat is left on the surface. The extraction is performed by means of ether, in a Soxhlet apparatus. It is essential that the paper be free from matters soluble in ether. A fat-free paper, manufactured by Schleicher & Schuell, is obtainable in strips suitable for the purpose. Each of these yields to ether from .001 to .002 gram of extract, and as this would only increase the figure for fat by .003 per cent., the error may be disregarded.

The procedure is as follows: 5 c. c. of the milk are discharged into a beaker 5 cm. high and 3.5 cm. in diameter. The charged beaker is weighed, and a strip of the paper which has been rolled into a coil, thrust into it. In a few minutes, the paper will absorb nearly the whole of the milk. The coil is then carefully withdrawn, and stood dry end downward on a sheet of glass.

With a little dexterity, all but the last fraction of a drop will be absorbed by the paper. The beaker is again weighed and the milk taken found by difference. It is of importance to take up the whole of the milk from the beaker, as the paper has a selective action, removing the watery constituents by preference over

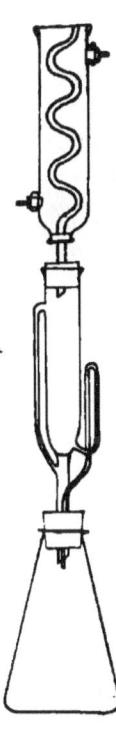
the fat. The charged paper is placed in the water-oven on a glass plate, milk end upwards, and dried. In about an hour it is usually in a suitable condition for the extraction of the fat. It is inserted in a Soxhlet continuous extraction apparatus, the tared flask of which should have a capacity of about 150 c.c. and contain about 75 c.c. of ether. Heat is applied to the flask by means of a water bath, or by resting it on a piece of asbestos paper, which is heated by a small flame. After the coil has received ten or twelve washings, the flask is detached, the ether removed by distillation and the fat dried by heating in an air oven, at about 220°F, and occasionally blowing air through the flask. After cooling, the flask is wiped with a piece of silk, allowed to stand ten minutes and weighed.

A convenient method of spreading the milk over the paper and arranging the coil is described by Allen. The strip is rolled up, together with a piece of thin string (previously boiled with water containing sodium carbonate to remove size and resinous matters) which serves to prevent contact between the concentric folds of the coil, and is conveniently passed through holes in the paper. A cap of filter paper is then placed over one end of the coil and secured by the ends of the string. The coil is then suspended by some simple means, the capped end being downward, and 5 c. c. of 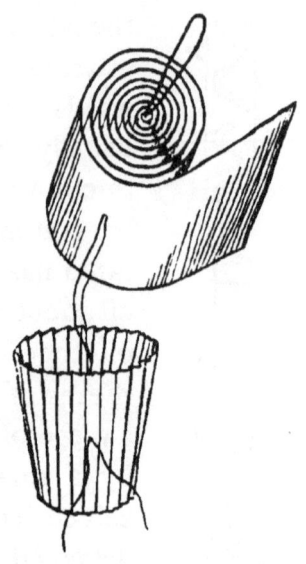 the milk to be tested, run on to the upper part from a pipette. The milk is rapidly absorbed by the paper and none filters through the cap. It is then dried in a water oven for an hour or two.

Sour milk must be weighed and thinned with a few drops of ammonium hydroxid before absorption by the paper.

Heavy filter paper of good quality, which yields to ether only a milligram or less, may also be employed, as suggested by Thomson. One end of the strip is fastened to some convenient support, and the other between the fingers. The paper being kept horizontal, 5 c.c. of the milk are taken in a pipette, and distributed evenly over it, the end of the pipette being wiped on a part left dry for the purpose. It is dried by passing to and fro a short distance over a Bunsen flame, or better, by suspending it in front of a stove.

Werner-Schmid Method.—This is a very satisfactory and rapid method for the determination of fat and is especially suitable for sour milk.

10 c. c. of the milk are measured into a long test-tube of 50 c. c. capacity, graduated to tenths c. c., and ten c. c. of strong hydrochloric acid added, or the milk may be weighed in a small beaker and washed into the tube with the acid. After mixing, the liquid is boiled one and a half minutes, or the tube may be corked and heated in the water bath from five to ten minutes, until the liquid turns dark brown. It must not be allowed to turn black. The tube and contents are cooled in water, 30 c. c. of *well washed* ether added, shaken, and allowed to stand until the line of acid and ether is distinct. The cork is taken out, and a double tube arrangement, like that of the ordinary wash-bottle, inserted. The stopper of this should be of cork and not of rubber, since it is difficult to slide the glass tube in rubber, and there is a possibility, also, of the ether acting on the rubber and dissolving it. The lower end of the exit tube is adjusted so as to rest immediately above the junction of the two liquids. The ethereal solution of fat is then blown

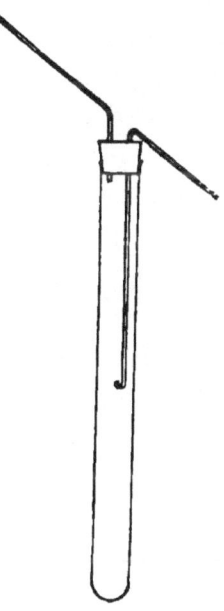

out, and received in a weighed flask. Two more portions of ether, 10 c.c. each, are shaken with the acid liquid, blown out and added to the first. The ether is then distilled off, and the fat dried and weighed as above.

Lactocrite Method.—In response to the demand for a method for fat determination, capable of affording fairly accurate results in the hands of dairymen and others unskilled in analytical operations, many processes have been devised, among which, those depending upon the use of centrifugal machines have proved most satisfactory. The lactocrite of De Laval is of this class. The following description of the machine and its use is given by H. Faber (*Analyst, XII, p. 6,*) "The apparatus itself consists of a strong round steel disc on a spindle, and test boxes of platinum-plated brass, with graduated glass tubes. 10 c. c. of the sample of milk to be tested are run into a small test glass, afterwards 10 c. c. of glacial acetic acid, containing 5 per cent. by volume of concentrated sulfuric acid, are run into the same glass, which is closed with a perforated cork-stopper, in which is inserted a piece of glass tube; this serves to prevent a concentration of the contents of the test glass during the boiling. In a water bath, arranged to hold twelve test glasses, these are heated by steam or gas for seven or eight minutes, after which time the casein has been completely dissolved, while the liquid has acquired a slight violet tinge. The next step is to charge the test boxes. These consist of a cup in which a perforated stopper fits tightly. The stopper holds the graduated glass tube. As the fat in the milk after boiling with acid has a great tendency to rise, the test-glass must be well shaken before its contents are poured into the cup, and when this is filled the stopper must be immediately pressed down in it whereby any excess of the mixture will escape through the glass tube, and the test-box is filled completely. After the test-boxes have been charged in this way, they are ready to be placed in the disc which will hold twelve at a time. The disc, which before use must be heated to about 110° F. by being placed in water of this temperature, has twelve cylindrical holes bored from a cavity on the top, radiating and a little sloping, In these the test-boxes are placed (if less than twelve test-boxes are used there should always be an even number placed so as not to disturb the equilibrium) and the cavity is filled with water, which will keep the liquid in the test-boxes from being pressed out by the centrifugal force. The disc, which fits any stand of a Laval separator, is now made to revolve for three or four minutes at ordinary speed (6000 revolutions in the minute). When it is again at rest, the test-boxes are drawn out and the column of fat on the graduated tube is read off, the divisions indicating immediately tenths per cent. of butter-fat by weight."

The results obtained with whole milks are very satisfactory, being within one-tenth per cent. and less, of those furnished by the Adams' process. With milks poor in fat, however, the results are low and it fails to indicate any fat whatever in milks which contain less than about .5 per cent. By substituting for the above mixture of sulfuric acid and acetic acid, one made of hydrochloric acid and lactic acid in the same porportion (1 to 19) this objection is overcome. The process so modified has been patented by the Separator Company of Stockholm.

Leffmann-Beam Method.—The cost of the lactocrite has prevented its general adoption. The following method devised by ourselves, can be applied much more cheaply both as regards the original cost of the apparatus, and of the chemicals required for the test— the latter being no inconsiderable item when many tests are made with the lactocrite. The manipulation is also very much simplified, the entire operation being performed in one test-bottle, and the use of the water-bath avoided.

The annexed cuts show the form of centrifugal machines, bottles and other accessories used. Machines arranged for either two, four, six, eight or twelve bottles are manufactured. The process is covered by pa-

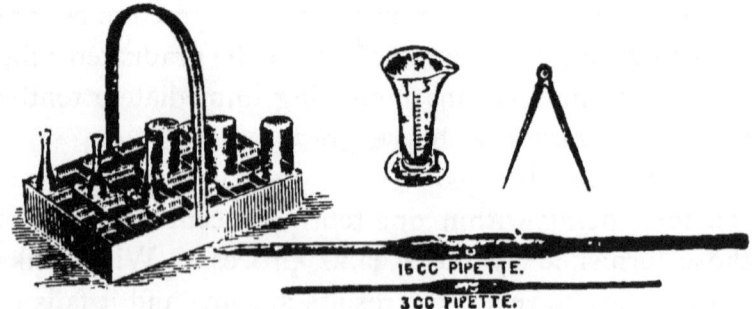

tent, in the United States, which has been assigned to J. E. Lonergan, of Philadelphia.

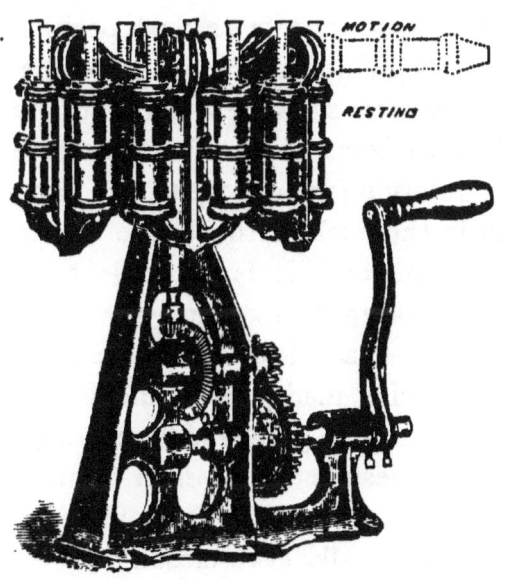

 The test-bottles, the form of which is shown in the cut, have a capacity of about 30 c. c., and are provided with a graduated neck, each division of which represents one-tenth per cent by weight of butter-fat.

15 c. c. of the milk are measured into the bottle, 3 c. c. of a mixture of equal parts of amyl alcohol and strong hydrochloric acid added, mixed, the bottle filled nearly to the neck with concentrated sulfuric acid and the liquids mixed by holding the bottle by the neck and giving it a gyratory motion. The neck is now filled to about

the zero point, with a mixture of sulfuric acid and water prepared at the time. It is then placed in the centrifugal machine, which is so arranged that when at rest the bottles are in a vertical position. If only one test is to be made, the equilibrium of the machine is maintained by means of a test-bottle, or bottles, filled with a mixture of equal parts of sulfuric acid and water. After rotation for from one to two minutes, the fat will collect in the neck of the bottle and the percentage may be read off. It is convenient to use a pair of dividers in making the reading. The legs of these are placed at the upper and lower limits respectively of the fat, allowance being made for the meniscus; one leg is then placed at the zero point and the reading made with the other. The results do not differ from those obtained by the Adams' process by more than one-tenth per cent., and are generally even closer.

Cream is to be diluted to exactly ten times its volume, the specific gravity taken, and the liquid treated as a milk. Since in the graduation of the test-bottles a specific gravity of 1030 is assumed, the reading must be increased in proportion.

A more accurate result may be obtained by weighing in the test-bottle about 2 c. c. of the cream and diluting to about 15 c. c. The reading obtained is to be multiplied by 15.45 and divided by the weight in grams of cream taken.

Calculation Method.—Several investigators have proposed formulæ by which when any two of the data, sp. gr., fat and total solids are known, the third can be calculated. Since the determination of the fat and specific gravity can be accurately made by rapid and simple methods, the formulæ become very servicable and should always be used to check the results. That of Hehner and Richmond is now exclusively used, having been based on extensive observation and perfect processes of fat extraction.

The formula is as follows:—

$$F = .859\,T - .2186\,G$$

in which F represents the fat, T the total solids and G the specific gravity. This formula will suffice for ordinary milks, but for poor skim milks it has been found necessary to modify it as follows:—

$$F = .859\,T - .2186\,G - .05\left(\frac{G}{T} - 2.5\right)$$

This correction is to be applied only when G divided by T exceeds 2.5.

In these formulæ, G represents the last two units of the specific gravity and any decimal. Thus, if the observed specific gravity be 1029.5, G will be 29.5.

Tables calculated from these formulæ will be found in the appendix so that by reference to the proper columns, the third datum may be obtained at once. A still more ready means of applying the formula is by the use of Richmond's slide-rule. This has three scales, two of which, for fat and total solids respectfully, are marked on the body of the rule, while that for the specific gravity is placed on the sliding portion.

The divisions are as follows:—

    Total solids . . . . 1 inch divided into tenths
    Fat . . . . . 1 164 inches " " "
    Specific Gravity . 0.254 " " " halves

The rule is employed by adjusting the arrow point to the graduation corresponding to the fat found, when the figure for the total solids will coincide with that for the observed specific gravity. This instrument does not take into consideration the necessary correction for poor skim milks, but the error from this cause will in no case exceed .08 per cent of fat and may usually be disregarded.

TOTAL PROTEIDS.

This determination is best made by calculation from the figure for total nitrogen obtained by Gunning's modification of Kjeldahl's process. The reagents and apparatus required are as follows:—

Standard Sulfuric Acid.—This should be about decinormal, the exact value being determined by precipitating a measured volume with barium chloride, collecting and weighing the barium sulfate under the usual precautions.

Standard Barium Hydroxid.—This should be about decinormal. The volumetric relation between this solution and the standard sulfuric acid must be accurately determined, employing phenolphthalein as an indicator.

Acid Potassium Sulfate Mixture.—One part of pure potassium sulfate is heated with two parts of sulfuric acid (strictly C. P.) until the potassium sulfate is dissolved. The mixture is semi-solid when cold but may be readily liquified by warming.

Sodium Hydroxid Solution.—A saturated solution in water. The best grades of pulverized caustic soda will suffice, such as those sold as concentrated lye for household use, *e. g.* "Banner Lye" and that of the Greenbank Alkali Co.

Digestion Flask.—A round-bottomed short-necked flask with a capacity of 300 c. c.

Distilling Flask.—A flask of ordinary shape, about 550 c. c. capacity, is fitted by means of a rubber stopper with a delivery tube, the lower end of which projects slightly below the stopper and should be cut off obliquely. The tube should be of the same diameter as that of the condensing tube noted below and should have one or two bulbs about four centimeters in diameter to prevent the carrying over of sodium hydroxid during the distillation.

Copper flasks are now manufactured for this purpose and will be found convenient and economical.

Condenser.—The condensing tube must be of block tin of an external diameter of about one centimeter. At least 30 centimeters of its length should be in contact with the cooling water. The junction of the glass and tin tube is made by a short close-fitting rubber tube, and the tubes are so bent as to slope towards the distilling flask. The lower end of the tin tube is connected by a short rubber tube with a glass bulb tube which dips into an Erlenmeyer flask of about 300 c. c capacity.

5 c. c. of the milk are weighed or measured into the flask and evaporated to dryness over the water-bath. 30 c.c. of the acid potassium sulfate mixture are added and heated over the Bunsen burner. At first, frothing occurs and white fumes escape, consisting chiefly of water vapor. To prevent loss of acid, the neck of the flask is now fitted with a funnel which is covered with a watch glass. This will cause the acid to condense and run back into the flask. The operation is finished when the liquid is colorless, which will generally be the case in about an hour. After cooling, the contents of the flask are transferred to the distilling flask by the aid of about 200 c. c. of water, and several pieces of ignited pumice dropped in. Sufficient of the sodium hydroxid solution (about 50 c. c.) is now added to make the mixture strongly alkaline. It should be poured down the side of the flask so that it does not mix at once with the acid. The flask is now connected with the condenser and the contents mixed by shaking. The Erlenmeyer flask into which the distillate is to be received, is charged with 20 c. c. of the standard sulfuric acid. The end of the delivery tube should dip below the level of this liquid. The liquid is now distilled until the whole of the ammonia is driven over.

TOTAL PROTEIDS. 33

Usually not more than 150 c. c. need be collected. The acid remaining unneutralized is determined by titration with the standard barium hydroxid, using phenolphthalein as an indicator, and the nitrogen distilled over as ammonia, calculated. The nitrogen figure multiplied by 6.38 will give that of the total proteids.

CASEIN AND ALBUMIN.

20 c. c. of the milk are mixed with 40 c. c. of a saturated solution of magnesium sulfate and powdered magnesium sulfate stirred in until no more will dissolve. After settling, the precipitate of casein and fat, including the trace of globulin, is filtered and washed several times with a saturated solution of magnesium sulfate. The filtrate and washings are saved for the determination of albumin. The filter and contents are transferred to a flask and the nitrogen determined by the method described above. The nitrogen so found multiplied by 6.38 gives the casein.

The filtrate and washings from the determination of casein are mixed, the albumin precipitated by a solution of tannin, filtered and the nitrogen in the precipitate determined as above. The same factor is used.

On account of the difficulty of washing the precipitated casein, we prefer to proceed as follows: 20 c. c. of the milk are mixed with saturated magnesium sulfate solution and the powdered salt added to saturation. The mixture is washed into a graduated measure with a small amount of the saturated solution, mixed, the volume noted, and allowed to stand until the separation takes place. The liquid is then filtered, as much as possible of the clear portion being drawn off with a pipette, and passed through the filter. An aliquot portion of the filtrate is taken, the albumin precipitated by a solution of tannin, and the nitrogen in the precipitate determined as above.

The casein is found by subtracting the figure for albumin from that for total proteids.

The Ritthausen Method for Total Proteids.—This method depends on the precipitation of the albuminoids by means of copper sulfate and sodium hydroxid. It is applicable only to fully developed milks; the proteids of colostrum are only partially precipitated. The following reagents are required.

Copper Sulfate Solution.—34.64 grams of pure crystalized copper sulfate are dissolved and made up to 500 c. c.

Sodium Hydroxid Solution.—About 12 grams are dissolved in 500 c. c. of water.

10 grams of the milk are placed in a beaker, diluted with 100 c. c. of distilled water, 5 c. c. of copper sulfate solution added and thoroughly mixed. The sodium hydroxid solution is then added drop by drop with constant stirring, until the precipitate settles quickly and the liquid is neutral, or at most very feebly acid. An excess of alkali will prevent the precipitation of some of the proteids.

TOTAL PROTEIDS.

The reaction should be tested on a drop of the clear liquid, withdrawing it by means of a rod, taking care not to include any solid particles. When the operation is correctly performed, the precipitate which includes the fat, settles quickly, and carries down all of the copper. It is washed by decantation with about 100 c. c. of water, and collected on a filter (previously dried at 265° F and weighed in a weighing bottle). The portions adhering to the sides of the beaker are dislodged with the aid of a rubber-tipped rod. The contents of the filter are washed with water until 250 c. c. are collected, which are mixed and reserved for the determination of the sugar as described below. The water in the precipitate is removed by washing once with strong alcohol, and the fat by six or eight washings with ether. The Soxhlet apparatus may be used for this purpose. The washings being received in a weighed flask, the determination of the fat may be made by evaporating the ether with the usual precautions.

The residue on the filter, which consists of the proteids in association with copper hydroxid, is washed with absolute alcohol which renders it more granular, and then dried at 265°F. in the air bath. It is weighed in a weighing bottle, transferred to a porcelain crucible, incinerated, and the residue again weighed. The weight of the filter and contents, less that of the filter and residue after ignition, gives the weight of the proteids. The results by this method are slightly high, owing to the fact that the copper hydroxid does not become completely converted into copper oxid at 265°

but the error so introduced is not great and may usually be disregarded.

SUGAR.

The following method due to Soxhlet, employs a Fehling's solution, made as required, by mixing equal parts of the following solutions:—

Copper Sulfate Solution.—34.64 grams of pure crystallized copper sulfate are dissolved in distilled water and made up to 500 c c.

Alkaline Tartrate Solution.—173 grams of pure sodium potassium tartrate, and 51 grams of sodium hydroxid of good quality, are dissolved and made up to 500 c. c.

100 c. c. of the mixed filtrate from the precipitated proteids are brought to boiling, in a beaker, 50 c. c. of boiling Fehling's solution added, and the boiling continued for six minutes. The precipitate is allowed to settle for a short time, and the supernatant liquid poured through a filter. About 50 c. c. of boiling water are added to the residue, and the heating continued for a minute or two. The precipitate is then conveyed to the filter, washed with boiling water, with alcohol and finally with a small quantity of ether. The filter and contents are dried in the water oven, the precipitate removed to a tared porcelain crucible, the filter held over the crucible and burnt to ash, which is added to the precipitate, and the cuprous oxid converted into cupric oxid by strong ignition for five or ten minutes over the Bunsen burner.

The amount of copper reduced under the conditions detailed above is not directly proportional to the milk sugar present. The following table shows the amounts of milk sugar ($C_{12}H_{22}O_{11} + H_2O$) equivalent to given weights of cupric oxid. The volumes of Fehling's solution and sugar solution, must conform strictly to the figures given above.

Cupric Oxid	Factor for 1 Milligram of Cupric Oxid	Milk Sugar	Cupric Oxid	Factor for 1 Milligram of Cupric Oxid	Milk Sugar
.4916	0.85	0.300	.3281	0.78	0.195
.4844	0.85	0.295	.3201	0.78	0.190
.4771	0.85	0.290	.3121	0.78	0.185
.4697	0.85	0.285	.3040	0.78	0.180
.4625	0.85	0.280	.2959	0.78	0.175
.4552	0.85	0.275	.2880	0.78	0.170
.4475	0.81	0.270	.2798	0.78	0.165
.4398	0.81	0.265	.2718	0.78	0.160
.4322	0.81	0.260	.2636	0.78	0.155
.4245	0.81	0.255	.2554	0.78	0.150
.4169	0.78	0.250	.2473	0.76	0.145
.4089	0.78	0.245	.2391	0.76	0.140
.4007	0.78	0.240	.2308	0.76	0.135
.3927	0.78	0.235	.2227	0.76	0.130
.3846	0.78	0.230	.2146	0.76	0.125
.3766	0.78	0.225	.2063	0.75	0.120
.3685	0.78	0.220	.1979	0.75	0.115
.3604	0.78	0.215	.1897	0.75	0.110
.3524	0.78	0.210	.1814	0.75	0.105
.3443	0.78	0.205	.1731	0.75	0.100
.3363	0.78	0.200			

The determination of sugar may also be made by means of the polarimeter after removal of the fat and proteids. This may be effected by means of a nitric acid solution of mercuric nitrate as suggested by Wiley.

The mercuric nitrate solution is prepared by dissolving mercury in an equal weight of nitric acid of

1.42 sp. gr. and adding to the solution an equal bulk of water.

60 c. c. of the milk are placed in a 100 c. c. flask and 1 c. c. of the mercuric solution added. The flask is filled to the mark with water, well shaken and the liquid filtered through a dry filter. The filtrate, which will be perfectly clear, may be examined in the polarimeter. Several readings should be made and the average taken.

It is to be noted that the actual volume of the sugar containing solution is 100 c. c., less the space occupied by the precipitated proteids and fat. The volume of fat is found by multiplying the weight in grams by 1.075 and the proteids by multiplying the weight by .8

For example:—

Sp. Gr. of milk 1030, Fat 4 per cent, Proteids 4 per cent.
Milk taken = 60 × 1.03 = 61.80 gms.
The weight of fat = 4 per cent of 61.80 = 2.47 gms.
The weight of proteids = 4 per cent. of 61.80 = 2.47 gms.
The volume of fat = 2.47 × 1.075 = 2.65 c. c.
The volume of proteids = 2.47 × .8 = 1.97 c. c.
The bulk of the sugar containing liquid is therefore
100 − (2.65 + 1.97) = 95.38 c. c.

In order to avoid the calculation involved in taking 60 c. c. of the milk as given above, an amount may be employed which is a simple multiple of the standard quantity to be used in the polarimeter at hand. Thus, for instruments adjusted so that 16.19 grams of sucrose (20.56 grams of milk sugar) in 100 c. c. of the

SUGAR. 39

solution produce a rotation of 100 degrees on the per cent. scale, 61.68 grams (20.56 × 3) may be weighed out directly for the purpose and made up to 100 c. c. plus the volume occupied by the fat and proteids, the latter being calculated as above. The sugar containing liquid will then be exactly 100 c. c., and the reading on the polarimeter divided by three will give the percentage of hydrated milk sugar direct, if a 200 mm. tube be employed. With a 400 mm. tube or 500 mm. tube the reading is to be divided by 6 or 7.5 respectively.

Polarimeters.—A discussion of the construction of the various forms of polarimeters and of the optical principles involved, would be beyond the scope of this work. It may be stated that the so called half-shadow instruments, for use with the sodium flame, are the most satisfactory. They are so arranged, by the use of a semicircle of thin quartz, that the field is divided into semicircles which are equally illuminated when the instrument registers zero. On the introduction of the tube carrying the sugar solution, the illumination becomes unequal and the angular rotation of the analyser which is required to restore the original condition, measures the rotation which has been caused by the sugar. Most instruments are furnished with two scales, one showing the rotation in angular degrees and the other expressing per cent. directly. The latter reads to 100 when a certain fixed quantity of the material has been dissolved in water and diluted to 100 c. c.

The *specific rotatory power* of a substance is the amount of rotation of the plane of polarized light, in angular degrees, produced by a solution containing one gram of the substance in one c. c., examined in a column one decimeter long.

It is expressed by the following formula in which

S is the specific rotatory power for light of wave length corresponding to the D line of the spectrum (sodium flame).

a is the angular rotation observed,

c is the concentration of the solution (weight in grams, in 100 c. c. of the liquid) and

l is the length of the tube in decimeters.

$$S = \frac{100\,a}{c \times l}$$

Calculation of the amount of sugar corresponding to the observed rotation may be made by substitution in the formula.

The sodium flame is most conveniently obtained by means of a bead of sodium carbonate, on platinum wire, heated in the flame of the Bunsen burner.

The specific rotatory power of milk sugar is unaffected by the concentration within the limits encountered in ordinary analytical work. It is slightly affected by temperature, being decreased by about .042 angular degree for each successive rise of one degree. The specific rotatory power at 68° F. is 52.5° when obseved by the sodium flame.

Birotation.—When freshly dissolved in cold water, milk sugar shows a higher rotation than that given above. By standing, or immediately on boiling, the rotatory power falls to the point mentioned. In preparing solutions from the solid milk sugar, care must be taken to bring them to the boiling point, previous to making up to a definite volume. This precaution is unnecessary when operating upon milk.

MILK ADULTERANTS.

Water.—The addition of water to milk is usually detected by the diminution in the amount of solids. Since nitrates are absent from normal milk, and almost invariably present in surface and subsoil waters, it may be possible to detect the addition of water by the application of one of the delicate tests for nitrates. The value of this method is considerably impaired by the fact that small amounts of nitrates may be introduced into the milk by the water used in rinsing the cans. Nevertheless, the presence of notable quantities of nitrates will be ground for grave suspicion. The test is applied as follows:—Several grains of diphenylamin are placed in a test-tube and dissolved in two or three c. c. of strong, pure sulfuric acid. A small quantity of milk is then added carefully, so as to form a layer on the surface of the acid. If nitrates be present, a blue color is formed at the junction of the two layers.

With any given sample of milk, addition of water decreases the gravity, while abstraction of fat increases it. It is possible, therefore, by carrying out both methods of adulteration carefully, to maintain the same gravity as in the original sample, so that this datum alone will not suffice to detect adulteration. Taken in conjuction with either the figure for fat or for total solids, the specific gravity becomes of direct value, and furnishes a means for determining, by calculation, the remaining datum.

For milk control in dairies, etc., it will suffice to take the specific gravity by the lactodensimeter (see page 16) and the fat by the Leffmann-Beam method. From the figures thus obtained the total solids can be ascertained by Hehner & Richmond's table (Table B) or Richmond's slide-rule.

Various substances are added to milk to conceal adulteration or inferiority in quality. The most frequently employed are coloring matters. Sugar, salt, starch and calf's brain have been added to milk, but are of infrequent occurence. It has occasionally been stated that chalk has been added, but this is obviously unlikely. The coloring matters most frequently employed are annatto, caramel, saffron, carotin and occasionally turmeric and certain coal-tar colors.

Annatto is easily detected by rendering the sample slightly alkaline by the addition of sodium acid carbonate, immersing in it a slip of filter paper and allowing it to remain over night. The presence of annatto will be indicated by a distinct reddish-yellow tinge to the paper.

Coal-tar colors are detected by adding to the milk ammonium hydroxid and allowing a piece of white wool to remain in it over night. The dye is taken up by the wool, which acquires a yellow tinge. When milk contains Martius' yellow, ammonium hydroxid intensifies the color and hydrochloric acid bleaches it.

Starch may be detected by the blue color developed on the addition of solution of iodin to the milk, which has previously been heated to the boiling temperature and then cooled. Starch is very often added to icecream and similar articles.

Salt and Cane sugar are occasionally added to milk that has been diluted with water. The former is easily detected by the taste, the increased proportion of ash and of chlorin. Cane sugar may be detected, if in considerable quantity, by the taste. The quantitive determination is made by the methods described in connection with condensed milk.

Additions such as calf's brain, may be detected by microscopic examination of the milk, or of the sediment, if any be formed on standing.

Antiseptic substances are now largely employed by dairymen and milk purveyors, especially in the warmer seasons, as a substitute for refrigeration. Preparations of boric acid and borax are most frequently used and are often sold under proprietary names which give no indication of their composition. Sodium carbonate is occasionally used to prevent coagulation resulting from slight souring.

Sodium Carbonate.—The following method, due to E. Schmidt, is stated to be capable of detecting onetenth of one per cent. of sodium carbonate, or of sodium acid carbonate.

10 c. c. of the milk are mixed with an equal volume of alcohol, and a few drops of a one per cent. solution of rosolic acid added. Pure milk shows merely a brownish-yellow color, but in the presence of sodium carbonate a more or less marked rose-red appears. The delicacy of the test is enhanced by making a comparison cylinder with the same amount of milk known to be pure. Any considerable addition of the salt may be detected by the increase in the ash, its marked alkalinity and effervescence with acid.

Benzoic Acid.—250–500 c. c. of the sample are rendered alkaline by a few drops of calcium or barium hydroxid, evaporated to one-fourth bulk, mixed with sufficient calcium sulfate to make a pasty mass and dried on the water bath. When condensed milk is examined, 100–150 grams should be mixed directly with sufficient calcium sulfate, and a few drops of barium hydroxid. Since the calcium sulfate is only employed to facilitate the drying, it may be replaced by powdered pumice or other inert material. The dry mass is finely powered, moistened with dilute sulfuric acid and then exhausted three or four times with about twice its volume of cold (50 per cent.) alcohol, which dissolves benzoic acid freely, but only mere traces of the fat. The alcoholic liquid which, in addition to the benzoic acid, contains milk sugar and mineral matter, is mixed thoroughly, neutralized with barium hydroxid and evaporated to small volume. The residue is acidified with weak sulfuric acid and extracted with successive small portions of ether. On evaporation, the ether leaves, almost pure, the benzoic acid which may be recognized by its odor and volatility.

Boric Acid.—When boric acid or borates are not present in quantities sufficient to appreciably increase the ash of the sample, the quantitive determination is difficult. The qualitative test is very delicate. 100 c. c. of the sample are rendered alkaline with calcium hydroxid, evaporated and ashed. Calcium hydroxid is preferred for this purpose because the ashing takes place more rapidly. The ash is dissolved in the smallest possible quantity of strong hydrochloric acid, the solution filtered and evaporated to dryness. An appreciable loss of boric acid will not occur. The residue is moistened with very dilute hydrochloric acid, mixed with tincture of turmeric and dried on the water bath. The smallest trace of boric acid gives to the residue a vermilion or cherry-red tint.

Concentrated hydrochloric acid gives with tincture of turmeric, a cherry-red color, which, however, disappears on addition of water and also becomes brown on drying, while the boric acid color appears only on drying and is not destroyed unless much water be added, or at the boiling point. The red color adheres strongly to the vessel and may be removed by alcohol. The flame test may be applied to the residue, but it is not delicate.

Quantitive determinations of boric acid may be made by Gooch's method as described in analytical manuals. Hehner has shown that sodium phosphate may be advantageously substituted for lime in this process, the details of the modification being given in *The Analyst* for August, 1891.

Salicylic Acid.—50 c. c. of the sample are treated with acid mercuric nitrate for the removal of the fat and proteids, as described in connection with the determination of milk sugar, and the liquid filtered. The filtrate is shaken violently with about one half its volume of a mixture of equal parts of ether and petroleum ether. The ethereal liquid is evaporated and a drop of neutral solution of ferric chlorid added to the residue. If salicylic acid be present, a characteristic violet color is developed. The reaction is very delicate.

DATA FOR MILK INSPECTION.

VARIATIONS IN COMPOSITION.

Average proportion of Solids in Milk.—The most extensive data on this point are those obtained by Vieth, a summary of whose results, covering a period of eleven years, was published in *The Analyst* of May, 1892. The total number of samples was 120,540, and although some changes had been made in the methods of analysis since the beginning of the work, the results were recalculated so as to be strictly comparable. The averages of the entire series were as follows:

Fat	4.1	per cent.
Non-fatty solids . .	8.8	"
Total solids . . .	12.9	"

Soxhlet states that normal cows' milk will contain at least

Total solids . .	12.00	per cent.
Non-fatty solids . .	8.50	"
Fat	3.50	"

Seasonal variations in the composition of milk.— The diagrammatic synopsis of the results obtained by Vieth, given on pages 48-49, being a reproduction of the table prepared by him, shows that a notable

48   Average Composition of 120, 540

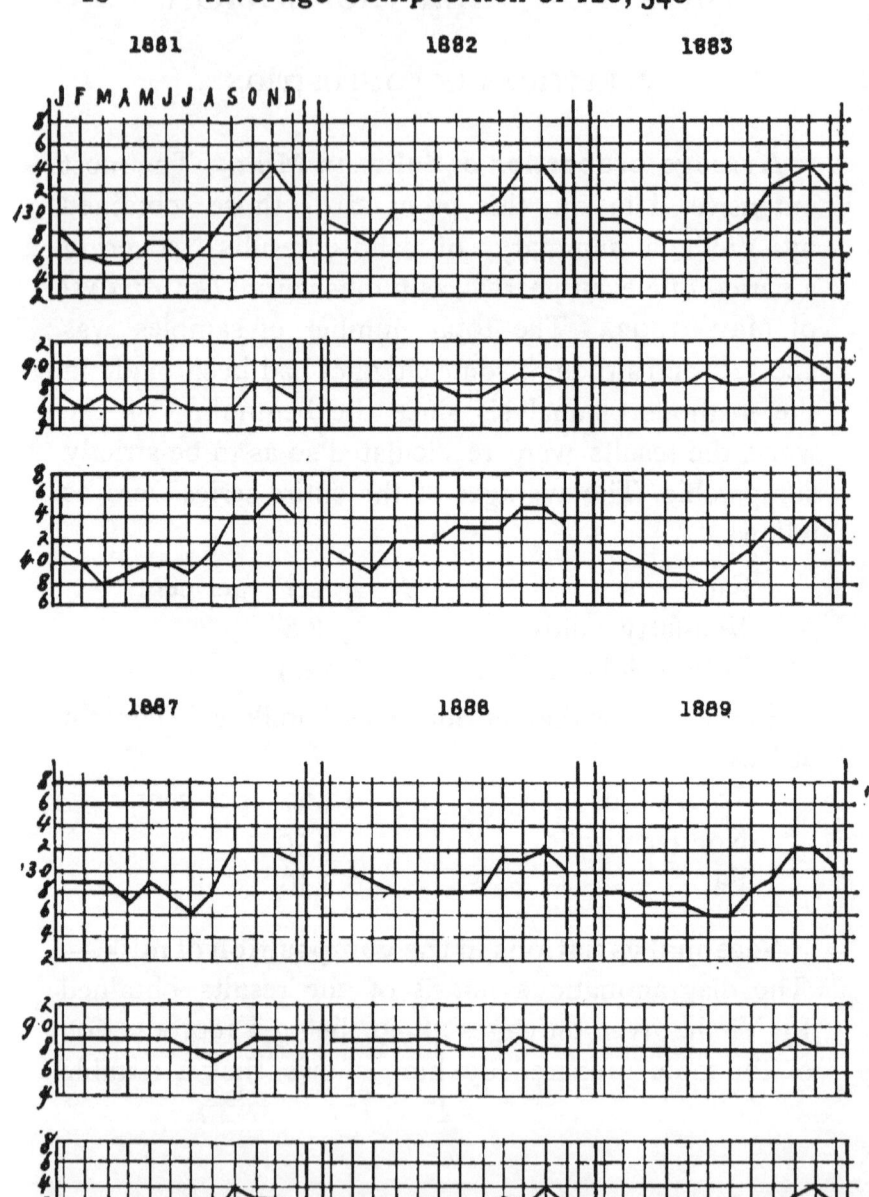

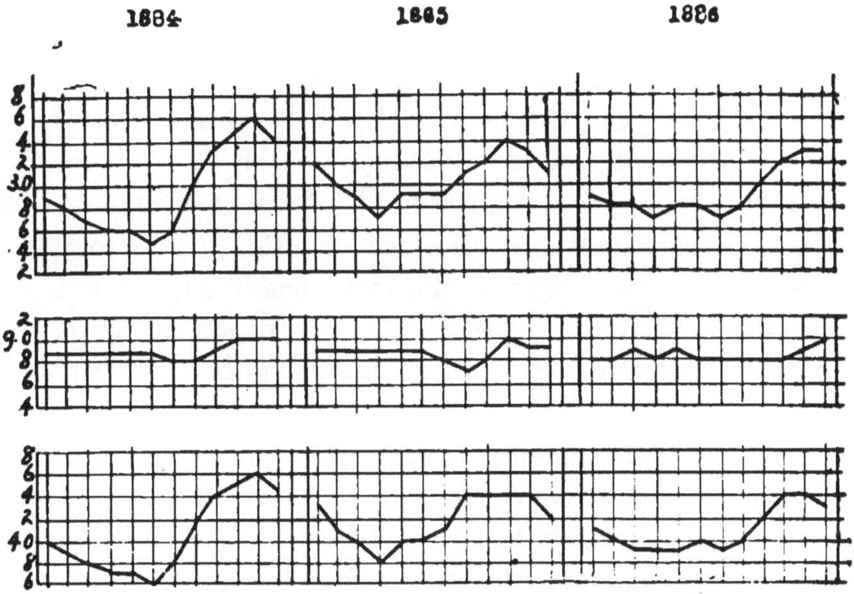

Samples of Milk---P. Vieth.

1884   1885   1886

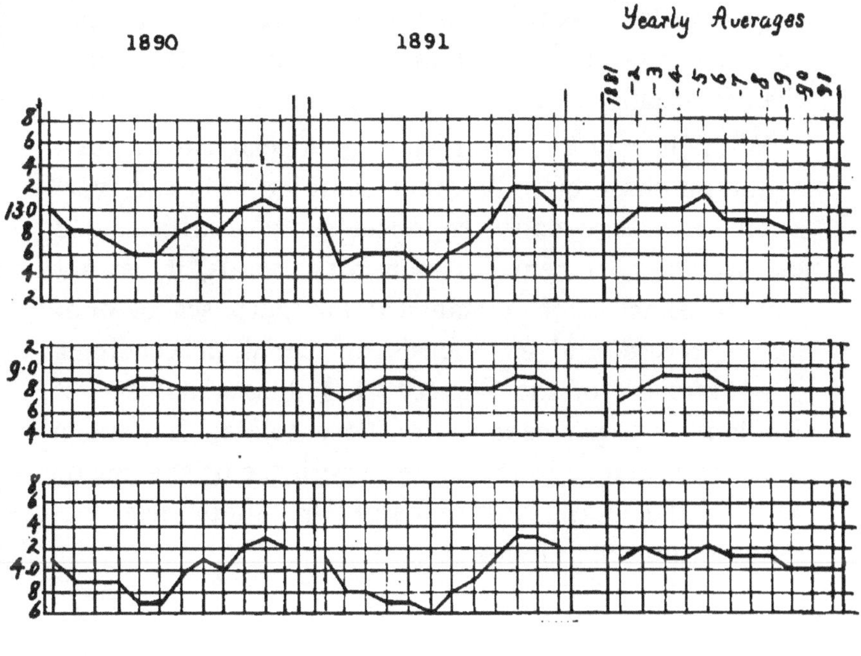

1890   1891   Yearly Averages

variation in the proportion of ingredients occurs during the year. The poorest quality occurs during the first half of the year, especially in April. A low figure is also frequently noted about July. In the fall the quality rises, being highest in October and November.

The diagram shows that the variations in the total solids are due mainly to the variations in fat, but not entirely, for an increase in the proportion of fat is usually attended by a slight increase in the non-fatty solids.

The earlier tendency was to assign too high a limit for the non-fatty solids, since this figure was obtained by methods which failed to extract all the fat. In applying, therefore, the more modern processes, normal milk will be found to yield a figure for the non-fatty solids decidedly below the extreme limit of 9.5 per cent. Even 9 per cent. for the non-fatty solids is more than is usually present.

While it may be permissible in special cases, such as the purchase of milk under contract, or in the operation of a large dairy, to reject samples which yield below nine per cent. of non-fatty solids, it is not just to exact such a standard for purposes of public inspection, and as a basis for penal proceedings. The standard of the Society of Public Analysts of England (8.5 per cent. of non-fatty solids), has been found satisfactory in the large experience of the members of that body, and recently Dr. Vieth has expressed himself as follows:

"My object is by no means to raise the cry that the standard adopted by the Society is too high; on the contrary, I think it is very judiciously fixed, but in upholding the standard of purity it should not be forgotten that the cows have never been asked for, nor given their assent to it, and that they will at times produce milk below standard. A bad season for haymaking is, in my experience, almost invariably followed by a particularly low depression in the quality of milk, toward the end of winter. Should the winter be of unusual severity and length, the depression will be still more marked. Long spells of cold and wet, as well as of heat and drought, during the time when cows are kept on pasture, also unfavorably influence the quality and, I may add, quantity of milk."

Deficient solids.—The following are some instances of deficiency of solids in milk known to be genuine.

SP. GR.	FAT	S. N. F.	T. S.	ANALYST
1029.6	3.38	7.95	11.33	C. B. Cochran
1030.0	3.62	8.31	11.93	"
1029.3	3.63	8.02	11.65	"
...	3.99	8.36	12.35	Leffmann & Beam
...	3.11	8.33	11.44	⎫ Monthly Averages
...	3.05	8.33	11.38	⎬ N. J. State Agric.
...	3.23	8.44	11.67	⎭ Exp. Station

In a herd of 60 cows, Richmond found 19 per cent. of the samples to contain between 8.38 and 8.50 per cent. solids not fat.

The following instances of unusually rich milk, were

reported in *The Analyst*, January, 1893.

Sp. Gr.	T'l Solids	Fat	Ash	Analyst
1026.6	19.50	11.06	.53	Smetham
1027.8	16.06	7.37	.72	"
1031.5	14.98	3.92	—	de Hailes

Since a partial creaming takes place in the udder, the first milkings (fore-milk) are poorer, and the last milkings (strippings) richer in fat, than the average milk. To insure a proper sample, the entire milking must be taken, as was done in the above analyses.

Variation according to breed.—The following figures taken from Bulletin 77, (1890) New Jersey State Agricultural Experiment Station, show the average composition of milk of various breeds of cattle:

AVERAGE COMPOSITION OF MILK FOR EIGHT MONTHS

Herd.	Specific Gravity	Percentage				
		Water	Total Solids	Fat	Casein	Sugar	Ash
Ayrshire	1034.1	87.30	12.70	3.68	3.48	4.84	0.69
Guernsey	1035.0	85.52	14.48	5.02	3.92	4.80	0.75
Holstein-Friesian	1032.8	87.88	12.12	3.51	3.28	4.69	0.64
Jersey	1035.3	85.66	14.34	4.78	3.96	4.85	0.75
Short-Horn	1033.9	87.55	12.45	3.65	3.27	4.80	0.73

Variations according to season.—The following table is condensed from the above report.

	Ayrshire		Holstein-Friesian		Jersey		Guernsey		Short-Horn	
	T. S.	Fat	T. S.	Fat.	T. S.	Fat.	T. S.	Fat.	T. S.	Fat.
Mar.	13.00	3.95	12.46	3.89	14.99	5.36	15.29	5.46	13.99	4.69
April	13.09	3.85	12.39	3.84	14.83	5.32	14.95	5.20	12.76	3.89
May	12.97	3.54	12.57	3.65	13.67	4.30	14.00	4.57	12.05	3.24
June	12.58	3.42	12.99	3.73	13.42	4.08	13.86	4.55	11.97	3.23
July	12.72	3.71	11.44	3.11	13.46	4.13	13.85	4.54	11.89	3.28
Aug.	13.08	4.07	11.38	3.05	13.60	4.22	13.93	4.81	12.08	3.56
Sept.	11.85	3.26	11.67	3.23	15.00	5.08	14.67	5.22	12.24	3.47
Oct.	12.27	3.60	12.08	3.55	15.75	5.71	15.28	5.78	12.61	3.82

Milk Standards.—Many efforts have been made to establish a minimum for the composition of normal milk, with a view to prevent adulteration. Standards proposed some years back, requiring a high proportion of non-fatty solids, were based upon analyses by methods which fail to extract the whole of the fat from the milk residue. The Society of Public Analysts of England formerly used a standard of 9 per cent. non-fatty solids and 2.5 per cent. fat, but when the improved method of analysis was adopted, altered the standard to the figures given in the table. The following are some of the standards which have been adopted:

DATA FOR MILK INSPECTION.

State, City, etc.	Percentage by Weights of Solids.		
	Non-fatty	Fat	Total
Pennsylvania, 1885, . . .	9.50	3.00	12.50
New York, 1884, . . .	9.00	3.00	12.00
New Jersey, 1882, . . .	9.00	3.00	12.00
Massachusetts, 1886, . . .	9.30	3.70	13.00
		May & June	12.00
Minnesota, 1889, ...	9.50	3.50	13.00
Columbus, Ohio, ...	9.375	3.125	12.50
Baltimore, Md., . , . .			12.00
Denver, Col., ...			12.00
Lansing, Mich., ...		3.00	12.50
Madison, Wis., ...		3.00	
Burlington, Vt., ...		3.50	12.50
Des Moines, Iowa, . . .		3.50	13.13
Portland, Oregon, . . .			12.00
Omaha, Nebraska, . . .		3.00	12.00
U. S. Treasury Department,	9.50	3.50	13.00
Philadelphia, 1890, Ordinance,	8.50	3.50	12.00
Society of Public Analysts of England, ...	8.50	3.00	11.50

SANITARY RELATIONS.

Although questions of the food value and the wholesomeness of milk are scarcely germain to this work, yet several practical points of comparatively recent developement are intimately connected with the work of the milk analyst and need some discussion. It is now well recognized that the dairy is an important factor in the distribution of disease, the influence taking place through several channels.

In the first place, dairy cattle are subject to several infectious diseases which are communicable to man and virulent in their effects. The most important of these is tuberculosis. The etiology of this affection is now clear. It is dependent on the development of a minute organism. The tendency at the present time is to the view that the keeping of dairy cattle is a fruitful cause of the spread of this disease. The specific germ of tuberculosis may be conveyed both in meat and milk and since the infection of the animal is not always recognised promptly, a most insidious source of danger exists. Other infectious and dangerous diseases *e. g.* scarlet fever, diphtheria and typhoid fever, may be conveyed by milk.

The common methods of adulterating milk, namely, by abstracting fat or adding water, diminish the food value but there has been great exaggeration of the importance of these changes. It can scarcely be sound to declare, as has occasionally been done by those engaged in promoting sanitary legislation, that milk reduced in fat by legitimate processes or even watered to a considerable extent is unwholesome. It is occasionally stated that the digestion of the proteids of milk is dependent on the presence of a certain amount of fat, but the experimental or clinical evidence of this is apparently not precise. Several competent authorities, *e. g.*, Vieth, Uffelmann and Hartshorne, have unhesitatingly declared even closely skimmed milk to be wholesome. As regards watered milk, it would be preposterous to assert that an article which is wholesome when containing nine per cent. of non-fatty solids, becomes unwholesome when containing eight per cent.

The unwholesomeness of milk arises not from change in the proportions of its principal ingredients, but from contamination with microorganisms. The danger from certain specific organisms has been mentioned, but the more frequent danger is from the ordinary non-pathogenic or putrefactive microbes, which, unless special care be taken, are invaribly present and multiply rapidly. To prevent such conditions, resort is had to sterilization by heat. Brief exposure to a temperature of $212°$ F is sufficient in most cases, but if the milk be subsequently exposed to air at ordinary temperatures, or mixed with unboiled water, it will be again contaminated and undergo putrefactive changes. In the warmer seasons of the year, these changes occur with great rapidity. Since clinical experience seems to show that boiled milk is frequently an unsatisfactory food for infants, methods of fractional sterilization at lower temperatures have been suggested. These depend on the fact, that, while spores and immature microbes require a rather high temperature for their destruction, fully developed organisms are more easily killed. By heating the milk, therefore, to a temperature much below the boiling point, the adult microbes are killed, while the milk solids are not unfavorably affected. The spores and immature organisms will, however, survive and may in a few hours develop, hence the milk is again heated, as before, and these later devepoped organisms will be killed. This process is repeated several times and finally complete sterilization is effected.

For the practical purpose of rendering milk safe as an article of food, it is not necessary to make repeated heatings. Numerous investigations are reported on this point, one of the most recent being a paper by Dr. R. G. Freeman (*Med. Rec.* June 10, 1893). A temperature of 167°F (75°C) continued for fifteen minutes, followed by rapid cooling by immersing the containing vessel in water, will kill the adult forms of most microbes, and milk so treated will remain unaltered for one or two days and will not have suffered any appreciable loss of digestibility, even for infants.

When it is considered that milk is almost the only form of animal food that is eaten in the uncooked condition, by civilized communities, the importance of the facts above noted will be apparent. Some interesting data as to the association of consumption, diphtheria, and similar diseases, with the maintenance of dairies have been collected, but the discussion of this feature of the question would be out of place here. Enough is known to show that raw milk is not a safe article of food, unless collected with such precautions as will prevent the introduction of infectious matter.

Artificial coloring matters do not involve any serious danger to health, except Martius' yellow, (dinitroalpnanaphthol) which is poisonous. The obvious objection to their use is that they enable milk of inferior quality to be substituted for rich milk. It is worthy of note that the assertion, occasionally made, that urine is employed in the preparation of annatto, is of little weight, since the annatto sold for diary use is prepared by unobjectionable methods.

Abnormal milks.—Milk occasionally becomes blue on the surface, the color forming in patches in proportion as the cream rises. The condition is due to the development of a chromogenic bacillus, first noted by Ehrenberg, and by him called *Vibrio syncyanus*, but now more correctly called *Bacillus syncyanus*. The condition sometimes prevails in epidemic form. The butter prepared from such milk possesses a greenish color and a disagreeable butyric odor. The bacillus seems to be non-pathogenic. Hueppe fed animals on food mixed with strong cultures of it, and observed no serious results. To prevent the development and spread of the bacillus it is recommended that the vessels intended to receive the milk be washed with boiling water. Reiset states that blue milk may be used for the production of butter by adding 0.5 gram of acetic acid to each liter (8 grains to the quart).

Red milk is due to accidental contamination with the *Bacillus prodigiosus*. The spores of this microbe exist in the atmosphere and rapidly develop when they fall upon any nutritive medium. The microbe does not appear to have any pathogenic properties.

Ropy milk.—This condition is occasionally seen during moist warm weather. The milk when drawn may not show any unusual properties, but in a few hours becomes so viscid that a spoonful of it may be lifted several inches without breaking the connection between the two portions. The nature and cause of the change are not known. The phenomenon generally appears rather suddenly and does not last long, almost always disappearing promptly on the advent of colder weather. Cases are known in which the milk thus affected has been used as food without any apparent unfavorable effect.

MILK PRODUCTS.

CONDENSED MILK.

A few brands of condensed milk in the market under the name of "evaporated cream," consist merely of whole milk concentrated to about two-fifths of its bulk, but most condensed milks contain a considerable amount of cane sugar. These samples represent, usually, whole milk concentrated to about one-third or two-sevenths of its original volume. On account of the presence of cane sugar, the fat is preferably estimated by the Adams' process, and for the same reason, in the determination of the total solids, the sample should be well spread in a thin layer to facilitate drying. The use of ignited asbestos in a platinum dish is to be recommended (see page 18).

Cane and milk sugars.—About 30 grams of the sample are accurately weighed, placed in a 100 c. c. flask, diluted to 80 c. c. and heated to boiling. The heating is necessary to avoid birotation, since condensed milk often contains crystallized milk sugar. The solution is cooled, 1.5 c. c. of acid mercuric nitrate solution (p. 37) added, the liquid made up to 100 c. c., well shaken, filtered through a dry filter and the polarimetric reading taken at once. It will be the sum of the effects of the two sugars. The volume of the sugar-containing liquid is calculated by allowing for the precipitated proteids and fat, as described under the determination of milk sugar.

50 c. c. of the filtrate are placed in a flask marked at 55 c. c., a piece of litmus paper dropped in and the excess of nitric acid cautiously neutralized by sodium hydroxid solution. The liquid is then faintly acidified by a single drop of acetic acid, (it must not be alkaline) a few drops of an alcoholic solution of thymol added, and then 2 c. c. of a solution of invertase, prepared by grinding half a cake of ordinary compressed yeast with 10 c. c. of water and filtering. The flask is corked and allowed to remain at a temperature of 100° to 110°F. for 24 hours. The cane sugar will be inverted while the milk sugar will be unaffected. The flask is filled to the mark (55 c. c.) with washed aluminum hydroxid and water, mixed, filtered and the polarimetric reading taken.

The rototary powers of cane sugar and dextrose are not appreciably affected by temperature within the limits of ordinary experiments. The same may be said of milk sugar (see page 40). Invert sugar, by reason of the levulose present, is materially affected by the temperature. Thus, a solution of cane sugar, which, before inversion, causes a rotation of $+$ 100 angular degrees, has after inversion, if observed at 32°F., a rotation of $-$44 degrees, a total change of 144; but at 70°F. the reading will be only—33 angular degrees, a total change of 133. The following formula is to be used for calculation.

$$C = \frac{100\,D}{144 - \frac{(t-32)}{3.6}}$$

in which C equals the angular rotation due to the uninverted cane sugar, D the difference in the polarimetric reading before and after inversion, and t the temperature in Fahrenheit degrees. Since, in the performance of the inversion, the liquid has been diluted from 50 to 55 c. c., the polarimetric reading must be increased in proportion, before the value of D is found. The value of C found by the equation, deducted from the reading before inversion, will give the angular rotation due to the milk sugar.

The specific rotatory power of cane sugar varies slightly with the concentration. Tollens gives the following formula, in which S is the sepecific rotatory power and C the concentration in grams per 100 c. c.

$$S = 66.386 + .015035\,C - .0003986\,C^2$$

A method which has given good results in our hands, is to take the reading for milk and cane sugars as described above and determine, in another portion of the sample, the milk sugar by Fehling's solution, which is not reduced by cane sugar. About three grams of the condensed milk (corresponding to about ten grams of ordinary milk) are accurately weighed, diluted with water, treated with copper sulfate and sodium hydroxid as described under the determination of proteids by the Ritthausen method, the liquid made up to 200 c.c., mixed, filtered through a dry filter and the reducing power of 100 c. c. of the filtrate determined by Soxhlet's method as described.

BUTTER.

Butter, commercially, consists of a variable mixture of fat, water and curd, obtained by churning cream from cow's milk. The water contains in solution milk sugar and the salts of the milk. Common salt is usually present, being added after the churning. Artificial coloring is frequently used.

The composition of commercial butter usually varies within the following limits:—

Fat	78 per cent.	to 94 per cent.
Curd	1 " "	" 3 " "
Water . .	5 " "	" 14 " "
Salt	0 " "	" 7 " "

Nostrums for butter making.—Preparations purporting to have the power to increase the yield of butter from a given weight of milk are now sold. One of these, advertised under the name "black pepsin," has been found to contain salt, annatto and a small amount of rennet. Pepsin has also been used. These curdle the milk and allow the incorporation of much cheese and water with the butter. It has been found that butter may also, without the addition of any chemicals, be incorporated with a large amount of cream.

Butter containing over forty per cent. of water, has been sold in this city. Such samples are pale and spongy, lose weight and become rancid very rapidly.

It is generally considered that butter should not contain more than 16 per cent. water. An excess of water diminishes the keeping quality.

The following methods for the analysis of butter have been adopted by the Association of Official Agricultural Chemists:—

"*Sampling.*—If large quantities of butter are to be sampled, a butter trier or sampler may be used. The portions drawn, about 500 grams, are to be carefully melted in a closed vessel, at as low a heat as possible, and when melted the whole is to be shaken violently for some minutes till homogeneous. The mass must be sufficiently solidified to prevent the separation of the water and fat. A portion is then placed in the vessel from which it is to be weighed for analysis, and should nearly or quite fill it. It should be kept in a cold place until analyzed. Determinations are made as follows:—

Water.—1.5–2.5 grams are dried to constant weight at the temperature of boiling water in a dish with a flat bottom, having a surface of at least 20 sq. cm.

Fat.—The dry butter from the water determination is dissolved in the dish with absolute ether, or with 76° benzin. The contents of the dish are then transferred to a Gooch crucible with the aid of a wash-bottle filled with the solvent, and are washed until free from fat. The crucible and contents are dried at the temperature of boiling water, until the weight is constant.

Casein and Ash.—The crucible containing the residue from the fat determination, consisting of the casein and salts, is covered and heated, gently at first, and gradually raising the temperature to just below redness. The cover may then be removed and the heat continued till the contents of the crucible are white. The loss in weight of the crucible and contents represents the weight of the casein, and the residue in the crucible, ash."

A weighed paper filter may replace the Gooch crucible.

Antiseptic substances in milk may find their way into the butter made from it. They will be dissolved in the water, and may be detected by separating this, by melting, and testing it as directed under milk.

Oleomargarin.—Under this term is now included by act of Congress, any oleaginous substance, intended as a substitute for butter, containing any proportion of fat other than butter-fat. The term "margarine" is employed in England, under authority of an act of Parliament, with the same significance. The principal materials employed in the preparation of butter-substitutes are cottonseed oil, mutton fat and beef fat. As usually manufactured, they are wholesome and economical.

Butter fat, like most fats, consists of a mixture of the ethers of tritenyl ($C_3 H_5$) but is peculiar among animal fats in containing notable proportions of acid radicles with a small number of carbon atoms. The exact arrangement is not known, but the weight of opinion is that it is not a mixture of simple fats, but that several acid radicles are united to the same tritenyl molecule. When saponified by sodium hydroxid and treated with acid, the individual fatty acids are obtained. It is upon the recognition of the peculiar acid radicles that the most satisfactory method of distinguishing butter from other fats is based. Since the relative proportion of these radicles differs in different samples, the quantitive estimation cannot be made with accuracy, but when the foreign fats are substituted to the extent of 25 per cent. or more, the adulteration can be detected with certainty and the quantitive determination approximately made.

The fatty acids containing a small number of carbon atoms, set free by the process of saponification and treatment with acid as noted above, are soluble in water and volatile. A method for their estimation depending on their solubilty in water was perfected by Hehner, but has now been displaced by a distillation method originally suggested by Hehner & Angell, but improved by Reichert and the details perfected by others, especially Wollny, and now generally known as the Reichert-Wollny method.

We have modified the process by substituting a solution of sodium hydroxid in glycerol as the saponifying agent, by which the time required is much shortened, the result subject to less variation, and the titration

BUTTER. 67

more exact. The following reagents are required.

Glycerol Soda.—100 grams of pure sodium hydroxid are dissolved in 100 c. c. of distilled water, and allowed to stand until clear. 20 c. c. of this solution are mixed with 180 c. c. of pure concentrated glycerol. The mixture may be conveniently kept in a capped bottle holding a 10 c. c. pipette.

Sulfuric Acid.—20 c. c. of pure concentrated sulfuric acid, made up with distilled water to 100 c. c.

Barium Hydroxid.—An approximately decinormal, accurately standardized, solution of barium hydroxid.

Indicator.—An alcoholic solution of phenolphthalein.

About 50 grams of the sample are placed in a beaker, and heated to a temperature of $120°$ to $140°F$, until the water and curd have settled to the bottom. The clear fat is then poured on a warm dry plaited filter, and kept in a warm place until 25 or 30 c. c. have been collected. If the filtrate is not perfectly clear, it should be reheated for a short time and again filtered.

A 300 c. c. flask is washed thoroughly, rinsed with alcohol and then with ether, and thoroughly dried by heating in the water oven. After cooling, it is allowed to stand for about 15 minutes and weighed. A pipette, graduated to 5.75 c. c., is heated to about $140°F$ and filled to the mark with the well mixed fat, which is then run into the flask. After standing for about fifteen minutes, the flask and contents are weighed. 20 c.c. of the glycerol soda are added and the flask heated over the Bunsen burner. The mixture may foam somewhat; this may be controlled, and the operation hastened, by shaking the flask. When all the water has been driven off, the liquid will cease to boil, and if the heat and agitation be continued for a few moments, complete saponification will be effected, the mixture becoming perfectly clear. The whole operation, exclusive of weighing the fat, requires less than five minutes. The flask is then withdrawn from the heat and the soap dissolved in 135 c. c. of water. The first por-

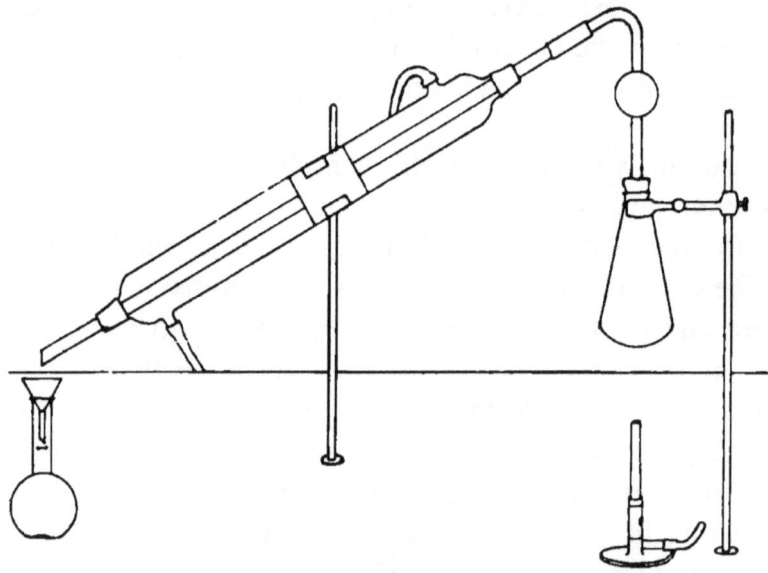

tions of water should be added drop by drop, and the flask shaken between each addition, in order to avoid foaming. When the soap is dissolved, 5 c. c. of the dilute sulfuric acid are added, a piece of pumice dropped in and the liquid distilled until 110 c. c. have been collected. The condensing tube should be of glass, and the distillation conducted at such a rate that the above amount of distillate is collected in 30 minutes.

The distillate is usually clear; if not, it should be thoroughly mixed, filtered through a dry filter, and 100 c. c. of the filtrate taken. To the distillate, about 0.5 c. c. of the phenolphthalein solution are added, and the standard barium hydroxid run in from a burette until a red color is produced. If only 100 c. c. of the distillate have been used for the titration, the number of c. c. of barium hydroxid should be increased by one-tenth.

When it is intended merely to distinguish pure butter and pure oleomargarin it will be sufficient to measure into the flask three or six c. c. of the clear fat, and operate upon this directly.

A blank experiment should be made to determine the amount of decinormal alkali required by the materials employed. With a good quality of glycerol, this will not usually exceed 0.5 c. c.

Butter (5 grams) yields a distillate requiring from 24 to 34 c. c. of decinormal alkali. Several instances have been published in which genuine butter has given a figure as low as 22.5 c. c., but such results are uncommon. The materials employed in the preparation of oleomargarin yield a distillate requiring less than 1 c. c. of alkali. Commercial oleomargarin is usually churned with milk in order to secure a butter flavor, and thus acquiring a small amount of butter-fat yields distillates capable of neutralising from 1 to 2 c. c. of alkali.

The method of determining the volatile acids of butter-fat, adopted by the Association of Official Agricultural Chemists, is as follows:—

Caustic soda solution.—100 grams of sodium hydroxid are dissolved in 100 c. c. of pure water. It should be as free as possible from carbonates, and be preserved from contact with the air.

Alcohol, of about 95 per cent., redistilled with caustic soda.

Acid.—Solution of sulfuric acid containing 25 c. c. of strongest acid in 1000 c. c. of water.

Barium hydroxid.—An accurately standardized, approximately decinormal solution of barium hydroxid.

Indicator.—Alcoholic solution of phenolphthalein.

Saponification flasks, 250 to 300 c. c. capacity, of hard well annealed glass, capable of resisting the tension of alcohol vapor at 212°F.

A pipette, graduated to deliver 40 c. c.

Distilling apparatus.

Burette.—A accurately calibrated burette, reading to tenths of a c. c.

Weighing the fat—The butter or fat to be examined should be melted, and kept in a dry warm place at about 145°F for two or three hours, until the moisture and curd have entirely settled out. The clear supernatant fat is poured off and filtered through dry filter paper in a jacketed funnel containing boiling water. Should the filtered fat in a fused state not be perfectly clear, the treatment above mentioned must be repeated.

The saponification flasks are prepared by having

BUTTER. 71

them throughly washed with water, alcohol and ether, wiped perfectly dry on the outside, and heated for one hour to boiling temperature. The flasks should then be placed in a tray by the side of the balance and covered with a silk handkerchief until they are perfectly cool. They must not be wiped with a silk handkerchief within fifteen or twenty minutes of the time they are weighed. The weight of the flasks having been accurately determined, they are charged with the melted fat in the following way:

A pipette with a long stem, marked to deliver 5.75 c. c. is warmed to a temperature of about 130°F. The fat having been poured back and forth once or twice into a dry beaker in order to throughly mix it, is taken up in the pipette and the nozzle of the pipette carried to near the bottom of the flask, having been previously wiped to remove any adhering fat, and 5.75 c. c. of fat are allowed to flow into the flask. After the flasks have been charged in this way, they should be recovered with the silk handkerchief and allowed to stand fifteen or twenty minutes, when they are again weighed.

The saponification.—Ten c. c. of 95 per cent. alcohol are added to the fat in the flask, and then 2 c. c. of the concentrated soda solution; a soft cork stopper is now inserted in the flask and tied down with a piece of twine. The saponification is then completed by placing the flask upon the water or steam bath. The flask during the saponification, which should last one hour, should be gently rotated from time to time, being careful not to project the soap for any distance up its sides. At the end of an hour the flask, after having been cooled to near the room temperature, is opened.

Removal of the alcohol.—The stoppers having been laid loosely in the mouth of the flask, the alcohol is removed by dipping the flask into a steam bath. The steam should cover the whole of the flask except the neck. After the alcohol is nearly removed, frothing may be noticed in the soap, and to avoid any loss from this cause or any creeping of the soap up the sides of the flask, it should be removed from the bath and shaken to and fro until the frothing disappears. The last traces of alcohol vapor may be removed from the flask by waving it briskly, mouth down, to and fro.

Dissolving the soap.—After the removal of the alcohol the soap should be dissolved by adding 100 c. c. of recently boiled distilled water, warming on the steam bath with occasional shaking, until solution of the soap is complete.

Setting free the fatty acids.—When the soap solution has cooled to about 145°F the fatty acids are separated by adding 40 c. c. of the dilute sulphuric acid solution mentioned above.

Melting the fatty acid emulsion.—The flask should now be re-stoppered as in the first instance, and the fatty acid emulsion melted by replacing the flask on the steam bath. According to the nature of the fat examined, the time required for the fusion of the fatty acid emulsions may vary from a few minutes to several hours.

BUTTER. 73

The distillation.—After the fatty acids are completely melted, which can be determined by their forming a transparent oily layer on the surface of the water, the flask is cooled to room temperature, and a few pieces of pumice stone added. The pumice stone is prepared by throwing it, at a white heat, into distilled water, and keeping it under water until used. The flask is now connected with a glass condenser, slowly heated with a naked flame until ebullition begins, and then the distillation continued by regulating the flame in such a way as to collect 110 c. c. of the distillate in, as nearly as possible, thirty minutes. The distillate should be received in a flask accurately graduated at 110 c. c.

Titration of the volatile acids.—The 110 c. c. of distillate, after thorough mixing, are filtered through dry filter paper and collected in a flask marked at 100 c. c. 100 c. c. of the filtered distillate are poured into a beaker holding from 200 to 250 c. c., 0.5 c. c of phenolphthalein solution added, and decinormal barium hydroxid run in until a red color is produced. The contents of the beaker are then returned to the measuring flask to remove any acid remaining therein, poured again into the beaker, and the titration continued until the red color produced remains apparently unchanged for two or three minutes. The number of cubic centimeters of decinormal barium hydroxid required should be increased by one-tenth.

Many other methods of detecting butter adulteration have been proposed. The specific gravity of the melted fat is of value in this connection, but the distinction between butter and its substitutes is not so sharp as with the distillation method. The same remark applies to the iodin absorption figure and to the refractive index. The latter can be accurately measured by means of an instrument called the oleo-refractometer.

Considerable use has been made of a method based upon the detection of crystalline structure by examination with polarized light. Such condition indicates however, merely that the sample has been previously melted. By churning oleo-oil with cream, a material is obtained which shows no crystalline structure when examined in this way.

Commercial forms of oleomargarin and butter exhibit characteristic differences on heating, which may be utilized for rapidly sorting a collection of samples. When butter is heated in a small tin dish directly over a gas flame, it melts quietly, foams and may run over the dish. Oleomargarin, under the same conditions, sputters noisily as soon as heated and foams but little. Even mixtures of butter and other fats show this sputtering action to a considerable extent. The effect depends upon the condition in which the admixed water exists, and the test is not applicable to butter which has been melted and reworked.

An alcoholic solution of sodium hydroxid heated for a moment with butter and then emptied into cold water, gives a distinct odor of pineapples (due to ethyl butyrate) while oleomargarin gives only the alcoholic odor.

Butter-Colors.—The following coloring matters are stated to have been found in butter : carrot color, annatto, marigold and carthamus flowers, turmeric and certain coal-tar colors.

The following test, described by E. W. Martin, we have found very satisfactory. Dissolve 2 parts of carbon disulfid in 15 parts of alcohol, by adding small portions of the disulfid to the alcohol and shaking gently; 25 c. c. of this mixture are placed in a convenient tube, 5 grams of the butter-fat added, and the tube shaken. The disulfid falls to the bottom of the tube, carrying with it the fatty matter, while any artificial coloring matter remains in the alcohol. The separation takes place in from one to three minutes. If the amount of the coloring matter is small more of the fat may be used.

CHEESE.

Cheese is obtained by the action of rennet, which is usually derived from the fourth stomach of the calf. The action is due to a non-organized ferment (enzyme) rennin, which acts directly on the proteids, and does not produce its effect through the intervention of acid. The curd (cheese) undergoes, by keeping, various decompositions, some essentially putrefactive, and undoubtedly due to the action of microbes. These changes, known as ripening, do not take place in a satisfactory manner when the curd is produced by the action of acids or neutral mineral salts, possibly because these agents interfere with the action of the microbes and enzymes on which the ripening depends. Skimmed milk is not infrequently used for the production of cheese, and foreign fats such as are used in the manufacture of oleomargarin are sometimes incorporated.

The analytical points to be determined in regard to cheese, are amounts of water, fat, casein, ash, the presence of fats other than butter-fat and coloring matters.

Care should be taken to select for analysis a sample which represents the average composition of the entire cheese. A thin section, reaching to the centre, is preferable, and portions from various parts of this should be cut fine and mixed. This should be done with as little exposure to air as possible, to avoid loss of water.

The following methods for water, fat, ash and casein are those provisionally adopted by the Association of Official Agricultural Chemists.

Water.—From 5 to 10 grams of cheese should be taken, and placed in thin slices in a weighed platinum or porcelain dish which contains a small quantity of freshly ignited asbestos, to absorb the fat which may run out of the cheese. The mass is then heated in a water oven for ten hours, and weighed; the loss in weight is to be considered as water.

Ash.—The dry residue from the water determination may be taken for the ash. If the cheese be rich, the asbestos will be saturated therewith. This mass may be ignited carefully, and the fat allowed to burn off, the asbestos acting as a wick. No extra heating should be applied during the operation, as there is danger of spurting. When the flame has died out, the burning may be completed in a muffle at low redness. When desired, the salt may be determined in the ash by titration with silver nitrate and potassium chromate.

Fat.—5 to 10 grams of the sample are ground in a small mortar with about twice the weight of anhydrous copper sulfate. The grinding should be continued until the cheese is finely pulverized and evenly distributed throughout the mass, which will have a uniform blue color. This mixture is transferred to a glass tube which has strong filter paper, supported by a piece of muslin, tied over the end. A little of the clean anhydrous copper sulfate is put into the tube next to the filter before introducing the mixture containing the cheese. On top of the mixture is placed a tuft of ignited asbestos, and the contents of the tube extracted with anyhydrous ether in the continuous extraction apparatus, for 15 hours. The ether is removed as usual and the fat dried at 212°F., to a constant weight.

This fat may be used to determine the presence or absence of oleomargarin, by applying the Reichert test.

Casein.—The nitrogen of about 2 grams of the cheese is determined by the Kjeldahl-Gunning method.

This multiplied by 6.38 gives casein.

The fat may be estimated by the centrifugal method, as follows:—

About 5 grams of the mixed cheese are weighed and transferred to the bottle, the last portions being washed in with the aid of water. A few drops of ammonium hydroxid are added, and sufficient water to make the liquid about 15 c. c. The liquid is warmed with occasional shaking, until the cheese is well disintegrated, and then treated as a sample of milk. The percentage of fat is found by multiplying the percentage reading by 15.45 and dividing by the number of grams of cheese taken for analysis.

Chrome yellow has been found in the rind of cheese. It may be detected by ashing the same in a porcelain crucible, assisting the burning of the carbon by a little nitric acid, and applying the usual tests for lead and chromium.

Cheese contains small amounts of milk sugar, lactic and other organic acids.

ANALYSES OF VARIOUS CHEESES.

	Proteids	Fat	Salts	Water
Cheshire.	36.1	25.5	4.8	30.4
Gruyère.	35.1	28.0	4.8	32.0
Roquefort	32.9	32.3	4.4	26.5
Cheddar	28.4	31.1	4.5	56.0
Camembert	18.9	21.0	4.7	51.9
Skim Milk	45.0	5.9	5.2	43.8

APPENDIX.

(A)

TABLE FOR CORRECTING THE SPECIFIC GRAVITY OF MILK ACCORDING TO TEMPERATURE.

BY DR. P. VIETH, F. C. S. &C.

DIRECTIONS FOR USE.

Find the Temperature of the Milk in the uppermost horizontal line, and the Specific Gravity in the first vertical column. In the same line with the latter, under the Temperature, is given the Corrected Specific Gravity.

For Example: Supposing the Temperature to be 51°, and the Specific Gravity, 34°, the Specific Gravity corrected to 60° Farenheit, is 32.9; or if the Temperature is 66°, and the Specific Gravity 29, the Corrected Specific Gravity is 29.8.

CORRECTIONS FOR TEMPERATURE.

Specific Gravity	DEGREES OF THERMOMETER (*Fahrenheit*)									
	46	47	48	49	50	51	52	53	54	55
1020	19.0	19.1	19.1	19.2	19.2	19.3	19.4	19.4	19.5	19.6
21	20.0	20.0	20.1	20.2	20.2	20.3	20.3	20.4	20.5	20.6
22	21.0	21.0	21.1	21.2	21.2	21.3	21.3	21.4	21.5	21.6
23	22.0	22.0	22.1	22.2	22.2	22.3	22.3	22.4	22.5	22.6
24	22.9	23.0	23.1	23.2	23.2	23.3	23.3	23.4	23.5	23.6
25	23.9	24.0	24.0	24.1	24.1	24.2	24.3	24.4	24.5	24.6
26	24.9	24.9	25.0	25.1	25.1	25.2	25.2	25.3	25.4	25.5
27	25.9	25.9	26.0	26.1	26.1	26.2	26.2	26.3	26.4	26.5
28	26.8	26.8	26.9	27.0	27.0	27.1	27.2	27.3	27.4	27.5
29	27.8	27.8	27.9	28.0	28.0	28.1	28.2	28.3	28.4	28.5
30	28.7	28.7	28.8	28.9	29.0	29.1	29.1	29.2	29.3	29.4
31	29.6	29.6	29.7	29.8	29.9	30.0	30.1	30.2	30.3	30.4
32	30.5	30.5	30.6	30.7	30.9	31.0	31.1	31.2	31.3	31.4
33	31.4	31.4	31.5	31.6	31.8	31.9	32.0	32.1	32.3	32.4
34	32.3	32.3	32.4	32.5	32.7	32.9	33.0	33.1	33.2	33.3
35	33.1	33.2	33.4	33.5	33.6	33.8	33.9	34.0	34.2	34.3

CORRECTIONS FOR TEMPERATURE.

Specific Gravity	DEGREES OF THERMOMETER (*Fahrenheit*)									
	56	57	58	59	60	61	62	63	64	65
1020	19.7	19.8	19.9	19.9	20.0	20.1	20.2	20.2	20.3	20.4
21	20.7	20.8	20.9	20.9	21.0	21.1	21.2	21.3	21.4	21.5
22	21.7	21.8	21.9	21.9	22.0	22.1	22.2	22.3	22.4	22.5
23	22.7	22.8	22.8	22.9	23.0	23.1	23.2	23.3	23.4	23.5
24	23.6	23.7	23.8	23.9	24.0	24.1	24.2	24.3	24.4	24.5
25	24.6	24.7	24.8	24.9	25.0	25.1	25.2	25.3	25.4	25.5
26	25.6	25.7	25.8	25.9	26.0	26.1	26.2	26.3	26.5	26.6
27	26.6	26.7	26.8	26.9	27.0	27.1	27.3	27.4	27.5	27.6
28	27.6	27.7	27.8	27.9	28.0	28.1	28.3	28.4	28.5	28.6
29	28.6	28.7	28.8	28.9	29.0	29.1	29.3	29.4	29.5	29.6
30	29.6	29.7	29.8	29.9	30.0	30.1	30.3	30.4	30.5	30.7
31	30.5	30.6	30.8	30.9	31.0	31.2	31.3	31.4	31.5	31.7
32	31.5	31.6	31.7	31.9	32.0	32.2	32.3	32.5	32.6	32.7
33	32.5	32.6	32.7	32.9	33.0	33.2	33.3	33.5	33.6	33.8
34	33.5	33.6	33.7	33.9	34.0	34.2	34.3	34.5	34.6	34.8
35	34.5	34.6	34.7	34.9	35.0	35.2	35.3	35.5	35.6	35.8

CORRECTIONS FOR TEMPERATURE.

Specific Gravity	DEGREES OF THERMOMETER (*Fahrenheit*)								
	66	67	68	69	70	71	72	73	74	75
1020	20.5	20.6	20.7	20.0	21.0	21.1	21.2	21.3	21.5	21.6
21	21.6	21.7	21.8	22.0	22.1	22.2	22.3	22.4	22.5	22.6
22	22.6	22.7	22.8	23.0	23.1	23.2	23.3	23.4	23.5	23.7
23	23.6	23.7	23.8	24.0	24.1	24.2	24.3	24.4	24.6	24.7
24	24.6	24.7	24.9	25.0	25.1	25.2	25.3	25.5	25.6	25.7
25	25.6	25.7	25.9	26.0	26.1	26.2	26.4	26.5	26.6	26.8
26	26.7	26.8	27.0	27.1	27.2	27.3	27.4	27.5	27.7	27.8
27	27.7	27.8	28.0	28.1	28.2	28.3	28.4	28.6	28.7	28.9
28	28.7	28.8	29.0	29.1	29.2	29.4	29.5	29.7	29.8	29.9
29	29.8	29.9	30.1	30.2	30.3	30.4	30.5	30.7	30.9	31.0
30	30.8	30.9	31.1	31.2	31.3	31.5	31.6	31.8	31.9	32.1
31	31.8	32.0	32.2	32.2	32.4	32.5	32.6	32.8	33.0	33.1
32	32.9	33.0	33.2	33.3	33.4	33.6	33.7	33.9	34.0	34.2
33	33.9	34.0	34.2	34.3	34.5	34.6	34.7	34.9	35.1	35.2
34	34.9	35.0	35.2	35.3	35.5	35.6	35.8	36.0	36.1	36.3
35	35.9	36.1	36.2	36.4	36.5	36.7	36.8	37.0	37.2	37.3

(B)

TOTAL SOLIDS CALCULATED FROM FAT AND SPECIFIC GRAVITY, BY THE FORMULA OF HEHNER & RICHMOND.

$$F = .859\, T - .2186\, G - .05 \left(\frac{G}{T} - 2.5 \right)$$
$$\text{if positive.}$$

Differences:
.01 Fat = .01 Total Solids.
.1 Sp. Gr. = .025 " "

Sp. Gr.	FAT									
	1	1.1	1.2	1.3	1.4	1.5	1.6	1.7	1.8	1.9
30.5	...	...	...	...	9.38	9.50	9.63	9.75	9.87	9.99
31.0	...	...	...	...	9.51	9.63	9.76	9.88	10.00	10.11
.5	...	...	...	...	9.64	9.77	9.89	10.01	10.13	10.25
32.0	...	...	...	...	9.77	9.90	10.02	10.14	10.26	10.37
.5	...	...	...	...	9.91	10.03	10.15	10.27	10.39	10.50
33.0	...	...	...	...	10.04	10.16	10.28	10.40	10.51	10.63
.5	9.71	9.83	9.95	10.06	10.17	10.30	10.41	10.52	10.64	10.76
34.0	9.84	9.96	10.08	10.19	10.30	10.43	10.54	10.65	10.78	10.88
.5	9.97	10.09	10.21	10.33	10.44	10.56	10.67	10.78	10.90	11.01
35.0	10.10	10.22	10.34	10.46	10.57	10.69	10.80	10.91	11.03	11.14

TOTAL SOLIDS CALCULATED.

Sp. Gr.	FAT									
	2	2.1	2.2	2.3	2.4	2.5	2.6	2.7	2.8	2.9
1024.0	8.42	8.54	8.66	8.77	8.89	9.01	9.12	9.24	9.35	9.47
.5	8.56	8.68	8.79	8.90	9.02	9.14	9.25	9.37	9.48	9.60
25.0	8.68	8.80	8.91	9.02	9.14	9.26	9.37	9.49	9.60	9.72
.5	8.81	8.93	9.04	9.15	9 27	9.39	9.50	9.62	9.73	9.85
26.0	8.94	9.06	9.17	9.28	9.40	9.52	9.63	9.75	9.86	9.98
.5	9.06	9.18	9.29	9 40	9.52	9.64	9.75	9.87	9.98	10.10
27.0	9.19	9.31	9.42	9.53	9.65	9.77	9.88	10.00	10.11	10.23
.5	9.32	9.44	9.55	9.66	9.78	9.90	10.01	10.13	10.24	10.36
28.0	9.45	9.57	9.68	9.79	9.91	10.03	10.14	10.26	10.37	10.49
.5	9.57	9.69	9.80	9 91	10.04	10.16	10.27	10.38	10.49	10.61
29.0	9.70	9.82	9.93	10.04	10.16	10.30	10.39	10.51	10.62	10.74
.5	9.83	9.95	10.06	10.18	10.29	10.42	10.52	10.64	10.75	10.87
30.0	9.96	10.09	10.20	10.31	10.42	10.54	10.65	10.77	10.88	11.00
.5	10.10	10.21	10.33	10.44	10.55	10.67	10.78	10.90	11.01	11.13
31.0	10.23	10.34	10.46	10.57	10.68	10.80	10.91	11.03	11.14	11.26
.5	10.36	10.47	10.58	10.69	10.80	10.92	11.03	11.15	11.26	11.38
32.0	10.49	10.60	10.71	10.83	10 93	11.05	11.16	11.28	11.39	11.51
.5	10.61	10.72	10.84	10.96	11.06	11.18	11.29	11.41	11.52	11.64
33.0	10.74	10.85	10.97	11.09	11.19	11.31	11.42	11.54	11.65	11.77
.5	10.87	10.98	11.09	11.21	11.32	11.44	11.55	11.66	11.77	11.89
34.0	11.00	11.11	11.22	11.34	11.44	11.56	11.67	11.79	11.90	12.02
.5	11.13	11.24	11.35	11.47	11.57	11.69	11.80	11.92	12.03	12.15
35.0	11.25	11.37	11.47	11.59	11.69	11.81	11.92	12.04	12.15	12.27

TOTAL SOLIDS CALCULATED.

Sp. Gr.	FAT									
	3.0	3.1	3.2	3.3	3.4	3.5	3.6	3.7	3.8	3.9
1024.0	9.59	9.70	9.82	9.94	10.05	10.17	10.29	10.40	10.52	10.63
.5	9.72	9.83	9.95	10.07	10.18	10.30	10.42	10.53	10.65	10.76
25.0	9.84	9.95	10.07	10.19	10.30	10.42	10.54	10.65	10.77	10.88
.5	9.97	10.08	10.20	10.32	10 43	10.55	10.67	10.78	10.90	11.01
26.0	10.10	10.21	10.33	10.45	10.56	10.68	10.80	10.91	11.03	11.14
.5	10.22	10.33	10.45	10 57	10.68	10.80	10.92	11.03	11.15	11.26
27.0	10.35	10.46	10 58	10.70	10.81	10.93	11.05	11.16	11.28	11.39
.5	10 48	10.59	10.71	10.83	10.94	11.06	11.18	11.29	11.41	11.52
28.0	10.61	10.72	10.84	10.96	11.07	11.19	11.31	11.42	11.54	11.65
.5	10.73	10.84	10.96	11 08	11.19	11.31	11.43	11.54	11.66	11.77
29.0	10.86	10.97	11.09	11.21	11.32	11.44	11.56	11.67	11.79	11.90
.5	10.99	11.10	11.22	11.34	11.45	11.57	11.69	11.80	11.92	12.03
30.0	11.12	11.23	11.35	11.47	11.58	11.70	11.82	11.93	12.05	12.16
.5	11.25	11.36	11.48	11.60	11.71	11.83	11.95	12.06	12.18	12.29
31.0	11.38	11.49	11.61	11.73	11.84	11.96	12.08	12.19	12.31	12.42
.5	11.50	11.61	11.73	11.85	11.96	12.08	12.20	12.31	12.43	12.54
32.0	11.63	11.74	11.86	11.98	12.09	12.21	12.33	12.44	12.56	12.67
.5	11.76	11.87	11.99	12.11	12.22	12.34	12.46	12.57	12.69	12.80
33.0	11.89	12.00	12.12	12.24	12.35	12.47	12.59	12.70	12.82	12.93
.5	12.01	12.12	12.24	12.36	12.47	12.59	12.71	12.82	12.94	13.05
34.0	12.14	12.25	12.37	12.49	12.60	12.72	12.84	12.95	13.07	13.18
.5	12.27	12.38	12.50	12.62	12.73	12.85	12.97	13.08	13.20	13.31
35.0	12.39	12.50	12.62	12.74	12 85	12.97	13.09	13.20	13.32	13.43

TOTAL SOLIDS CALCULATED.

Sp. Gr.	FAT									
	4.0	4.1	4.2	4.3	4.4	4.5	4.6	4.7	4.8	4.9
1024.0	10.75	10.87	10.98	11.10	11.22	11.33	11.45	11.57	11.68	11.80
.5	10.88	11.00	11.11	11.23	11.35	11.46	11.58	11.70	11.81	11.93
25.0	11.00	11.12	11.23	11.35	11.47	11.58	11.70	11.82	11.93	12.05
.5	11.13	11.25	11.36	11.48	11.60	11.71	11.83	11.95	12.06	12.17
26.0	11.26	11.38	11.49	11.61	11.73	11.84	11.96	12.08	12.19	12.30
.5	11.38	11.50	11.61	11.73	11.85	11.96	12.08	12.20	12.31	12.42
27.0	11.51	11.63	11.74	11.86	11.98	12.09	12.21	12.33	12.44	12.55
.5	11.64	11.76	11.87	11.99	12.11	12.22	12.34	12.46	12.57	12.68
28.0	11.77	11.89	12.00	12.12	12.24	12.35	12.47	12.59	12.70	12.81
.5	11.89	12.01	12.12	12.24	12.36	12.47	12.59	12.71	12.82	12.93
29.0	12.02	12.14	12.25	12.37	12.49	12.60	12.72	12.84	12.95	13.06
.5	12.15	12.27	12.38	12.50	12.61	12.73	12.85	12.97	13.08	13.19
30.0	12.28	12.40	12.51	12.63	12.74	12.86	12.98	13.10	13.21	13.32
.5	12.41	12.53	12.64	12.76	12.87	12.99	13.11	13.23	13.34	13.45
31.0	12.54	12.66	12.77	12.89	13.00	13.12	13.24	13.36	13.47	13.58
.5	12.66	12.78	12.89	13.01	13.12	13.24	13.36	13.48	13.59	13.70
32.0	12.79	12.91	13.02	13.14	13.25	13.37	13.49	13.61	13.72	13.83
.5	12.92	13.04	13.15	13.27	13.38	13.50	13.62	13.74	13.85	13.96
33.0	13.05	13.17	13.28	13.40	13.51	13.63	13.75	13.87	13.98	14.09
.5	13.17	13.29	13.40	13.52	13.63	13.75	13.87	13.99	14.10	14.21
34.0	13.30	13.42	13.53	13.65	13.76	13.88	14.00	14.12	14.23	14.34
.5	13.43	13.55	13.66	13.78	13.89	14.01	14.13	14.25	14.36	14.47
35.0	13.55	13.67	13.78	13.90	14.01	14.13	14.25	14.37	14.48	14.59

TOTAL SOLIDS CALCULATED.

Sp. Gr.	FAT									
	5.0	5.1	5.2	5.3	5.4	5.5	5.6	5.7	5.8	5.9
1024.0	11.91	12.03	12.15	12.26	12.38	12.50	12.61	12.73	12.85	12.96
.5	12.04	12.16	12.28	12.39	12.51	12.63	12.74	12.86	12.98	13.09
25.0	12.16	12.28	12.40	12.51	12.63	12.75	12.86	12.98	13.10	13.21
.5	12.29	12.41	12.53	12.64	12.76	12.88	12.99	13.11	13.23	13.34
26.0	12.42	12.54	12.66	12.77	12.89	13.01	13.12	13.24	13.36	13.47
.5	12.54	12.66	12.78	12 89	13.01	13.13	13.24	13.36	13.48	13.59
27.0	12.67	12.79	12.91	13.02	13.14	13.26	13.37	13.49	13.61	13.72
.5	12.80	12.92	13.04	13.15	13.27	13.39	13.50	13.62	13.74	13.85
28.0	12 93	13.05	13.17	13.28	13.40	13.52	13.63	13.75	13.87	13.98
.5	13.05	13.17	13.29	13 40	13.52	13.64	13.75	13.87	13.99	14.10
29.0	13.18	13.30	13.42	13.53	13.65	13.77	13.88	14.00	14.12	14.23
.5	13.31	13.43	13.55	13.66	13.78	13.90	14.01	14.13	14.25	14.36
30.0	13.44	13.56	13.68	13.79	13.91	14.03	14.14	14.26	14.38	14.49
.5	13.57	13.69	13.81	13.92	14.04	14.16	14.27	14.39	14.51	14.62
31.0	13.70	13.82	13.94	14.05	14.17	14.29	14.40	14.52	14.64	14.75
.5	13.82	13.94	14.06	14.17	14.29	14.41	14.52	14.64	14.76	14.87
32.0	13.95	14.07	14.19	14.30	14.42	14.54	14.65	14.77	14.89	15.00
.5	14.08	14.20	14.32	14.43	14.55	14.67	14.78	14.90	15.02	15.13
33.0	14.21	14.33	14.45	14.56	14.68	14.80	14.91	15.03	15.15	15.26
.5	14.33	14.45	14.57	14.68	14.80	14.92	15.03	15.15	15.27	15.38
34.0	14.46	14.58	14.70	14.81	14.93	15.05	15.16	15.28	15.40	15.51
.5	14.59	14.71	14.83	14.94	15.06	15.18	15.29	15.41	15.53	15.64
35.0	14.71	14.83	14.95	15.06	15.18	15.30	15.41	15.53	15.65	15.76

TOTAL SOLIDS CALCULATED.

Sp. Gr.	FAT								
	6.0	6.1	6.2	6.3	6.4	6.5	6.6	6.7	6.8	6.9
1024.0	13.08	13.20	13.31	13.42	13.54	13.66	13.78	13.90	14.01	14.13
.5	13.21	13.33	13.44	13.55	13.67	13.79	13.91	14.03	14.14	14.26
25.0	13.33	13.45	13.56	13.67	13.79	13.91	14.03	14.15	14.26	14.38
.5	13.46	13.58	13.69	13.80	13.92	14.04	14.16	14.28	14.39	14.51
26.0	13.59	13.71	13.82	13.93	14.05	14.17	14.29	14.41	14.52	14.64
.5	13.72	13.83	13.94	14.05	14.17	14.29	14.41	14.53	14.64	14.76
27.0	13.84	13.96	14.07	14.18	14.30	14.42	14.54	14.66	14.77	14.89
.5	13.97	14.09	14.20	14.31	14.43	14.55	14.67	14.79	14.90	15.02
28.0	14.10	14.22	14.33	14.44	14.56	14.68	14.80	14.92	15.03	15.15
.5	14.22	14.34	14.45	14.56	14.68	14.80	14.92	15.04	15.15	15.27
29.0	14.35	14.47	14.58	14.69	14.81	14.93	15.05	15.17	15.28	15.40
.5	14.48	14.60	14.71	14.82	14.94	15.06	15.18	15.30	15.41	15.53
30.0	14.61	14.73	14.84	14.95	15.07	15.19	15.31	15.43	15.54	15.66
.5	14.74	14.86	14.97	15.08	15.20	15.32	15.44	15.56	15.67	15.79
31.0	14.87	14.99	15.10	15.21	15.33	15.45	15.57	15.69	15.80	15.92
.5	14.99	15.11	15.22	15.33	15.45	15.57	15.69	15.81	15.92	16.04
32.0	15.12	15.24	15.35	15.46	15.58	15.70	15.82	15.94	16.05	16.17
.5	15.25	15.37	15.48	15.59	15.71	15.83	15.95	16.07	16.18	16.30
33.0	15.38	15.50	15.61	15.72	15.84	15.96	16.08	16.20	16.31	16.43
.5	15.50	15.62	15.73	15.84	15.96	16.08	16.20	16.32	16.43	16.55
34.0	15.63	15.75	15.86	15.97	16.09	16.21	16.33	16.45	16.56	16.68
.5	15.76	15.88	15.99	16.10	16.22	16.33	16.46	16.58	16.69	16.81
35.0	15.88	16.00	16.11	16.22	16.34	16.45	16.58	16.70	16.81	16.92

As already noted on page 13, the heating of milk causes an alteration in the milk sugar. In *The Analyst* for June and July, 1893, H. D. Richmond and L. K. Bosely note that heating to the extent to which milk is subjected in the preparation of condensed milk may reduce the rotatory power of the sugar sufficiently to cause serious error, if the polarimeter be used for the determination. The reducing power with Fehling's solution is not seriously affected.

In the examinations made by us, we have not noticed so great an alteration in the milk sugar, but in view of the liability to error from this cause, the cane sugar is best determined by the difference in polarimetric reading produced by inversion, (pp 60–62) and the milk sugar gravimetrically (p 62).

The albumin of condensed milk is partly coagulated by the heat employed in its manufacture. When magnesium sulfate is added, therefore, to precipitate the casein, the coagulated albumin will be carried down at the same time and only the soluble albumin will be found in the filtrate. Faber (*Analyst*, 1889) has applied this fact to the detection of the previous heating of a sample of milk. Usually, only about one-third of the albumin is found uncoagulated in condensed milk. The soluble albumin of unheated milk ranges from 0.35 to 0.50 per cent.

Calf's brain in milk, 43
Cane-sugar, 43, 61
Casein in milk, 10, 33
—— — butter, 65
—— — cheese, 77

KJELDAHL–GUNNING Method, 30

LACTOCRITE, 24
Lactodensimeter, 16

INDEX.

ABNORMAL milks, 51, 52
   Adam's method, 20
Adulterants, 41
Albumin, 10
———, determination of, 33
Amphoteric reaction, 12
Annatto, 42
Antiseptics, 43
Ash of butter, 65
——— cheese, 76
——— milk, 10, 19
Average of solids, 47

BACTERIA in milk, 12
   Benzoic acid, 44
Birotation, 40
Black pepsin, 63
Boiling, effects of, 13
Borax, 43, 45
Boric acid, 43, 45
Butter, 63
———, analysis of, 64
——— colors, 75
——— substitutes, 65
Buttermilk, 14

CALCIUM phosphate in milk, 10
Calculation method, 28
Calf's brain in milk, 43
Cane-sugar, 43, 61
Casein in milk, 10, 33
——— butter, 65
——— cheese, 77

Caseinogen, 10
Cheese, 75, 78
Citric acid, 10
Coal-tar colors, 42
Colostrum, 11
Condensed milk, 60
Cream, 14
———, evaporated, 60

DATA for milk inspection, 47
   Decomposition of milk, 13
De Leval Method, 24
Diphenylamin test, 41
Diseases conveyed by milk, 45

ENZYMES in milk, 10
   Evaporated cream, 60

FAT in butter, 64
   Fat in cheese, 77
Fat in milk, 20
Fehling's solution, 36
Formula for fat, 29
Freezing, effect of, 13

GLOBULIN in milk, 10
   Glycerol-soda, 67

HEHNER & RICHMOND'S formula, 29

KJELDAHL–GUNNING Method, 30

LACTOCRITE, 24
   Lactodensimeter, 16

INDEX.

Lactose, 9
——, determination of, 36, 41
Leffmann-Beam method, 26

MARGARINE, 65
　Mercuric nitrate solution, 38
Microbes in milk, 12, 56
Milk control, 42
——, coagulation of, 10
——, formation of, 9
—— globules, 9
——, properties of, 12
—— proteids, of, 10
—— standards, 15, 51
—— sugar, 9
——, composition of, 11

NITRATES in milk, 41
　Nostrums for butter making, 63

OLEOMARGARIN 65

PEPSIN, black, 63
　Polarimeters, 39
Poor milk, causes of, 51
Proteids, 30
Pyknometer, 17

REICHERT-WOLLNY Method, 66
Richmond slide rule, 29
Ritthausen Method, 34

SALICYLIC acid, 46
　Salt, 43
Sanitary relations, 54
Seasonal variations in milk, 41, 47, 48, 50, 51, 53

Skimmed milk, 55
Skimming, 15
—— ——, centrifugal, 14
Sodium carbonate, 43
Society of Public Analysts' standard, 50, 53
Solids, deficient, 51
Solids, excessive, 52
Spirillum choleræ in milk, 12
Soxhlet, composition of milk, 47
Specific gravity, 15
—— —— bottle, 17
—— —— of milk, change in, 12
Standards, official, 54.
Starch, 43
Sterilization, 12, 56
—— ——, fractional, 56
Sugar, cane, 43, 61
——, milk, 9, 36

TOTAL solids, 17
　Turmeric, 42

VARIATIONS in milk, 47, 54
　Vieth on milk standards, 51
Vieth, table of averages, 47
Volatile acids in butter, 66
—— ——, Leffmann-Beam method for, 66
—— ——, official method for, 70

WATER in butter, 64
　Water in cheese, 76
Water in milk, 41
Werner-Schmid method, 23
Westphal balance, 16

www.ingramcontent.com/pod-product-compliance
Lightning Source LLC
Chambersburg PA
CBHW032242080426
42735CB00008B/972